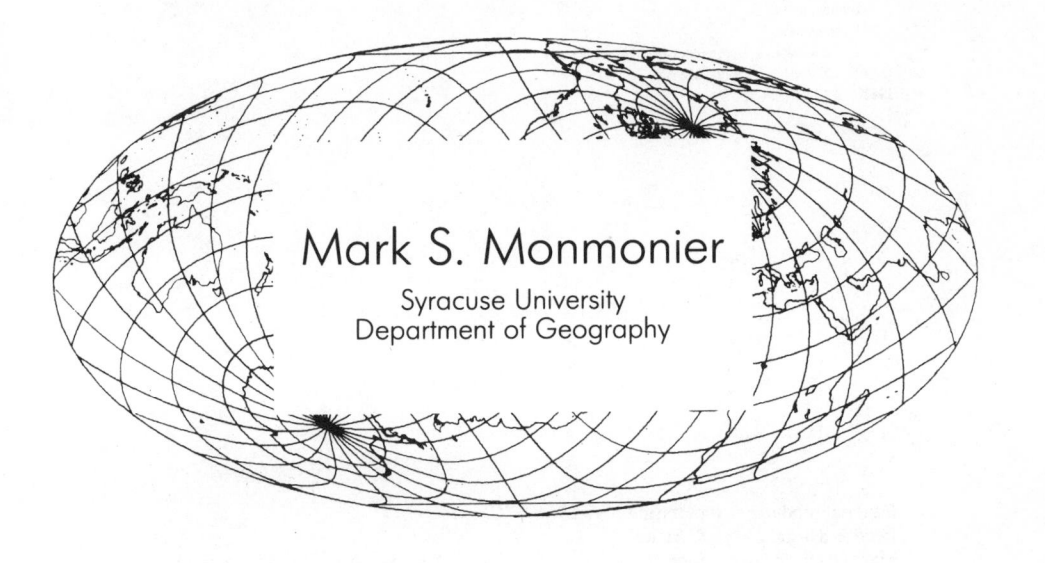

Mark S. Monmonier

Syracuse University
Department of Geography

COMPUTER-ASSISTED CARTOGRAPHY

PRINCIPLES AND PROSPECTS

Prentice-Hall, Inc., Englewood Cliffs, N.J. 07632

Library of Congress Cataloging in Publication Data

MONMONIER, MARK S.
  Computer-assisted cartography.

  Bibliography: p.
  Includes index.
  1. Cartography—Data processing. I. Title.
GA102.4.E4M66   526'.028'54   81-14380
ISBN 0-13-165308-3   AACR2

Editorial/production supervision: *Gretchen K. Chenenko*
Interior design: *Leslie I. Nadell*
Manufacturing buyer: *John Hall*
Cover idea: *Michelle Kermes*
Cover design: *Carol Zawislak*

Printed in the United States of America
10 9 8 7 6 5 4 3 2 1

ISBN 0-13-165308-3

Prentice-Hall International, Inc., *London*
Prentice-Hall of Australia Pty. Limited, *Sydney*
Prentice-Hall of Canada, Ltd., *Toronto*
Prentice-Hall of India Private Limited, *New Delhi*
Prentice-Hall of Japan, Inc., *Tokyo*
Prentice-Hall of Southeast Asia Pte. Ltd., *Singapore*
Whitehall Books Limited, *Wellington, New Zealand*

TO GEORGE F. DEASY (1912–1977)

Intelligence expert during World War II in the Department of State and the topographic branch of the Department of the Army, Professor of Geography at the Pennsylvania State University until his retirement in 1973, and a guiding but not stifling mentor to many.

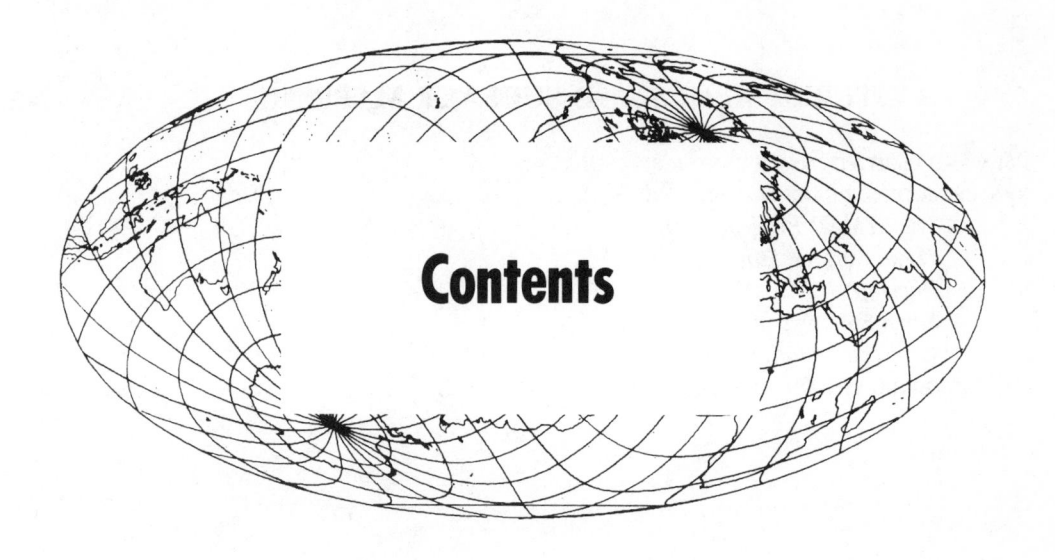

Contents

3

RASTER SYMBOLS AND SURFACE MAPPING 45

4

RASTER-MODE MEASUREMENT AND ANALYSIS 67

5

VECTOR SYMBOLS 89

6

CARTOMETRY AND MAP PROJECTIONS 113

7

CARTOGRAPHIC DATA STRUCTURES 137

8

COMPUTER-ASSISTED MAP DESIGN 157

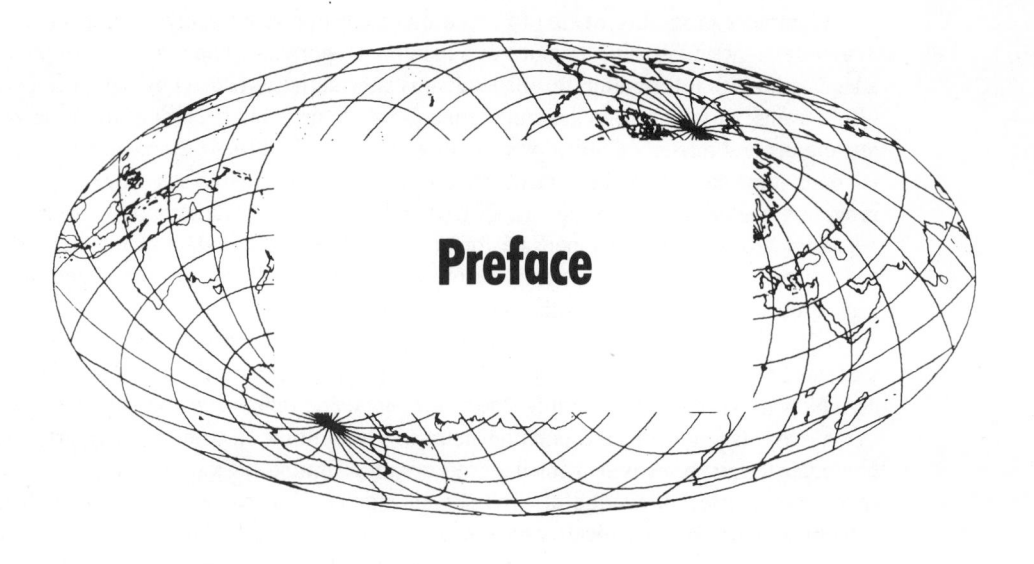

Preface

This book is about computers and mapping: how computers can make maps, and how computers eventually will change both the nature of mapping and the appearance of maps. The effects of electronic data processing and communications equipment upon maps and atlases will be as profound as its effects upon finance, manufacturing, warfare, and entertainment. Using the digital computer as a process control unit for cartographic drafting and as an accounting machine for geographic information is a special kind of computer application, and one that only recently has emerged from a stage of exploratory development to the more mature arena of fine-tuning.

In examining this technological innovation, the book looks only briefly at hardware for capturing and displaying map data to avoid encumbering the reader with minor details. Rather, emphasis is upon an understanding of the principles whereby present and future automated systems store, retrieve, analyze, and display geographic data. The first two chapters introduce the computer-assisted capture and display of data, and explain how computers, programs, and data files can serve as archives for massive amounts of map information and plot tailor-made maps in seconds. The next four chapters explore the relationships among map symbols, automated map analysis, and the two primary types of digital map data, raster and vector. Although numerous examples are presented, ranging from how a computer draws a circle to how smoothed contour lines can be obtained from randomly scattered elevation points, the book is not a textbook on computer programming. Chapter Seven addresses the importance of organizing data files to fulfill various mapping objectives, and the final chapter provides several examples of how the computer assists the map designer in a thorough analysis, before plotting, of the data and range of graphic possibilities.

Computer programs written to make maps are likewise largely ignored in an effort to avoid the inevitable obsolescence of a handbook approach. The likelihood of providing adequate how-to information for the student is substantially reduced by the great diversity of programs, display terminals, and computing systems available today. Both the student and the general reader should appreciate that the longer half life of fundamental principles is more important than the ephemeral usefulness of detailed facts. The two mapping programs treated here in any detail, CMAP and SYMAP, are sufficiently widespread and pedagogically useful to compensate for being well behind the state of the art. Specific, up-to-date information on a wider variety of systems and data for automated mapping can be obtained by writing to the organizations listed in the Appendix.

The book is designed especially for geographers and cartographers, the primary students of thematic mapping, but it is hoped that geologists, foresters, planners, landscape architects, environmental scientists, demographers, and other users of computer-produced maps may also find the presentation helpful. Definitions of common cartographic and computer terms are provided for the reader with a limited background in either of these two areas. Some knowledge of high school algebra and trigonometry will be helpful. Despite a basic, nontechnical approach, the book should also prove interesting and informative to computer scientists, civil engineers, and land surveyors.

I am indebted to many colleagues who offered suggestions and encouragement, especially Susan Moorlag, Eric Anderson, and my other part-time associates in the Division of National Mapping of the U.S. Geological Survey; Don Meinig and John Thompson in the Department of Geography at Syracuse University, who convinced me that writing a book is not an awesome task once you decide to go ahead and do it; and Jim Carter, of the University of Tennessee, Tony Williams of Pennsylvania State University, and Bob Wittick of Michigan State University, who provided many helpful comments on an earlier draft. Most of the illustrations were prepared in the Syracuse University Cartographic Laboratory by Doug Flewelling, Marcia Harrington, Anne Perry, Donald O'Keefe, Michelle Kermes, Eric Lindstrom, and Chris McNerney, under the supervision of Mike Kirchoff. The manuscript was typed ever so carefully by Joyce Bell, and the galleys received thorough additional readings by Andy Douglas and Tim Petersen. At Prentice-Hall, Betsy Perry, Leslie Nadell, Logan Campbell, Chuck Durang, Gretchen Chenenko, and David Stirling contributed experience and guidance during various stages of the project. My wife Marge was most understanding of my mildly eccentric working habits during preparation of the manuscript.

M.S.M.

Syracuse, NY

1

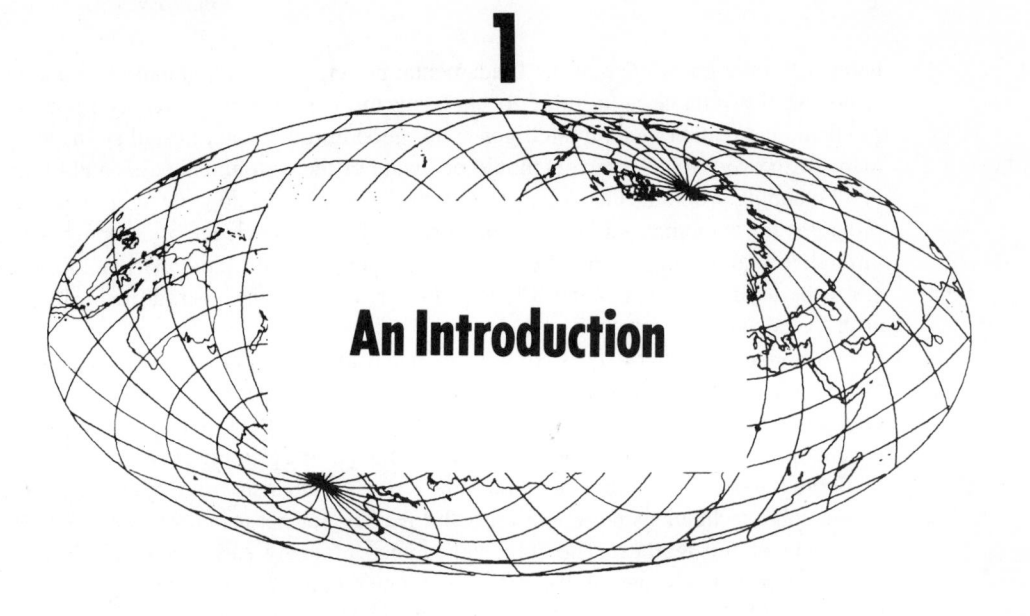

An Introduction

The computer can increase the value of the map as a decision-making tool. Maps provide important information for significant decisions at many levels. Families use maps to plan vacations. Armies use maps to organize military assaults. People use maps to locate the homes of new friends. Companies use maps to find locations for new factories. Environmental groups and planners use maps to evaluate the possibly adverse effects of proposed shopping malls, landfills, and generating stations. Decisions based on outdated or erroneous information lead to personal or societal complications. Yet many maps are out of date because of the time and cost required to compile and edit new maps or revise old ones. The computer is fast and precise, and has a useful role to play in the production of maps.

However strong your interest in using computer-produced maps in solving problems or generating ideas, you might have general reservations about computers and their not entirely positive influence on human life. Public concern about possible invasions of privacy has followed the development of large collections of easily retrieved personal data, and the excuse "computer error" is used all too commonly to shift responsibility from inept management and unqualified employees to a mindless and guileless piece of electronic machinery. To be sure, the computer can amplify the consequences of both malevolence and sloppiness, but neither fear nor skepticism can mount a convincing argument against using computers to make maps and increase our useful supply of environmental information.

The marriage of mapping and the computer does raise a legitimate suspicion about the validity of maps produced at low cost and in great numbers by well-meaning computer users with little knowledge of the fundamental principles of cartography. Common sense is not wholly reliable as a guide to choosing the best mapping strategy for a specific problem, and neither general intelligence nor good intuition can forestall an inappropriate map. At the very minimum you should be aware of the cartographic uses and limitations of all mapping programs that you are likely to use. If you plan to write your own computer programs for mapping, you ought to acquire a working knowledge of mapping principles through formal instruction, private study, or competent advice, particularly if the programs are to be used by others. After all, you do not want to join others in using "computer error" as a shield.

Automated cartography is, of course, much more than the mere linking of sophisticated process-control equipment to the traditional mapping methods of the 1950s. The digital computer has had a profound effect on maps, an effect that will equal or surpass the changes in mapping occasioned by the invention of the printing press and the discovery of photography. Cost, time, and possible cries of anguish from traditionally trained cartographic technicians no longer are valid reasons for not using methods once deemed inefficient or too tedious. The stark appearance of many early computer-drawn maps reflects a primitive state of the art in computer cartography rather than an inherent limitation. Moreover, map communication can now be more personal, as well as visually appealing: the technology that litters John Smith's mailbox with advertisements promising fulfillment to "the entire Smith family" if they subscribe to a magazine can with equal dispatch produce a useful travel atlas for "The Smith Family's Summer Trip to Maine."

The computer has changed not only mapping as a process but also the map as a concept, and the cartographer who ignores the potential of the computer is as naive and misguided as the mapping programmer or program user who ignores cartographic theory. The computer revolution produced orbiting satellites that have expanded land-cover mapping to the entire globe and improved remote-sensing systems that extend the cartographer's vision well beyond the small part of the electromagnetic spectrum visible to the human eye. Maps from satellite imagery might resemble photographs, televised pictures, or conventional maps. The ephemeral map, presented on the cathode-ray tube of an interactive graphics system, with the perspective, the symbolization, and even the theme readily altered at the command of the operator, has changed greatly the physical form of some maps, as well as the nature of map use. Moreover, maps are no longer designed solely for humans; computers can extract geographic information from special machine-readable maps called *digital geographic data bases* and even analyze this information to propose solutions to problems. In effect, the computer is also a map user.

This chapter introduces the subject of computer-assisted cartography by looking first at computer-controlled machines for graphic display. The original impetus for making maps by computer came largely from computer scientists and manufacturers of computing machinery. The ongoing development of new and more flexible display devices promises significant changes in the content and appearance of maps. The second section of the chapter is a brief historical perspective on the effect upon cartography of earlier technological revolutions. One notable change brought about by the computer is the paramount need to plan carefully for the orderly, efficient implementation of geographic data bases and display systems. The final section of the chapter discusses some fundamental issues in implementation that must be addressed by persons or agencies intending to use computers to generate their maps.

COMPUTER HARDWARE FOR GRAPHIC DISPLAY

A discussion of the electrical and mechanical devices used to display maps is a natural starting point for a text on computer cartography. The recognition that machines able to perform accurately the thousands of tedious, repetitive calculations associated with map projections and land surveys might be more useful if the results were presented in a rough geographic form provided the early impetus for computer-assisted cartography. The primitive computer-drawn maps of the 1950s were crude, coarse drawings produced on electric typewriters, which were the principal means of communication between the early electronic computer and its operator.

The basic format of the typewriter map persisted well after high-speed line printers relieved much of the bottleneck of computed results that could be generated much faster than they could be displayed. Line-printer maps consist of letters and numbers positioned within cells organized as an array of horizontal rows and vertical columns, and usually elongated vertically. Resolution was low, but almost every computer had a line printer, and every programmer could write and test programs with graphic displays as readily as programs with only tabular displays. Maps produced on line printers were rarely published, except as illustrations in reports extolling the benefits of automated mapping. Line-printer maps are still used in research, more so at the important but less conspicuous stages of idea generation and informal probing than as convincing and more formal expository graphics. Developments in both graphic *hardware,* the machines that display maps, and cartographic *software,* the programs and data files used to generate the maps, have progressed far beyond someone's inevitable idea that mapping on a typewriter, if rapid and ready, might indeed be useful.[1]

Typewriter mapping has survived not only in the efficient but crude line-printer maps of the researcher but also in the basic data structure for numerous display devices with a far finer resolution. Early recognition of the cartographic potential of the television screen was as predestined as that for the line printer: the computer logic controlling the carriage and selecting the characters for an electric typewriter can be modified to control the position and intensity of a beam of electrons striking the phosphorescent inner coating of a television picture tube (Fig. 1-1). The pattern of horizontal lines scanned by this electron beam, or cathode ray, is called a *raster.* Cartographic information organized in raster mode can also be recorded as a map image on standard paper with ink, on sensitized paper with heat or electric current, and on photographic film with the coherent light of

---

[1]So great is the complexity of graphic display hardware and the diversity of designs used by different manufacturers, that a comprehensive introduction in this chapter is impracticable. The discussion that follows focuses upon salient features of particular interest to the cartographer and attempts to explore what might be done, rather than precisely how data are displayed and captured. These basic principles are likely to survive the important but volatile techniques of display hardware.

Figure 1-1 Raster-mode color display terminal developed for the Decision Information Display System (DIDS) by NASA has a separate CRT (right) for displaying alphanumeric information and a joystick (center) for referencing locations on the color CRT (left). (Courtesy NASA.)

a laser beam. Raster-mode display devices start at the top of a picture and proceed downward; hence the map can be described by a series of specifications indicating how many lines farther down, if any, and how far across the last line the electron beam should proceed before changing its intensity. This one-line-after-another organization is compatible with data stored on magnetic tape, essentially a long, thin, sequential memory, divided into *records* that can be arranged to correspond to the scan lines of a raster display (Fig. 1-2).

Raster organization is particularly useful for map data scanned from existing maps by a *drum scanner* or from the land below by a scanner in an orbiting satellite or a high-altitude reconnaissance plane. A printed map can be mounted on a drum rotating rapidly beneath an optical sensor that moves slowly along the length of the drum. The image is divided into scan lines, which in turn are divided into *pixels,* discrete picture elements. For each pixel the sensing head measures the intensity of reflected light and converts this analog intensity to a digital response on, for example, a scale from 1 for very light to

128 for very dark. The digital intensities are recorded on magnetic tape for further processing (Fig. 1-2).

Scanning the Earth from a space satellite is different only in scale, not in concept: the satellite moves over the Earth in a scan track divided into parallel scan paths perpendicular to the direction of the track (Fig. 1-3). A small oscillating mirror directs the radiation reflected from the scan paths to a sensing head that performs the analog-to-digital, light-to-number conversion for the pixels into which the scan paths are divided. The digital responses are then transmitted to ground recording stations and stored on magnetic tape.

Cartographic information can also be stored as *vector* data, with linear features such as rivers recorded as the *(X, Y)* coordinate pairs of pivot points linking short straight-line segments. Features are thus encoded as sequential lists of point coordinates, together with any additional descriptive codes needed. Geographic areas such as census tracts and counties can be represented as *area polygons,* closed chains of straight-line segments bounding administrative regions. Electromechanical *digitizers,* with which human operators follow lineations and boundaries and record point coordinates at specified intervals or notable pivot points, are a principal means of *capturing* data in vector mode (Fig. 1-4). Automatic line-following digitizers, which sense and track linear features with servomechanisms, can also be used, but an operator must usually be present to resolve uncertainties presented by intersecting features.

Resolution in vector mode often is determined by the precision with which point coordinates may be measured and recorded. A digitizer recording only to the nearest half-millimeter cannot detect as much fine detail as a digitizer recording to the nearest tenth of a millimeter. Whereas resolution can be measured for raster data as the size of the

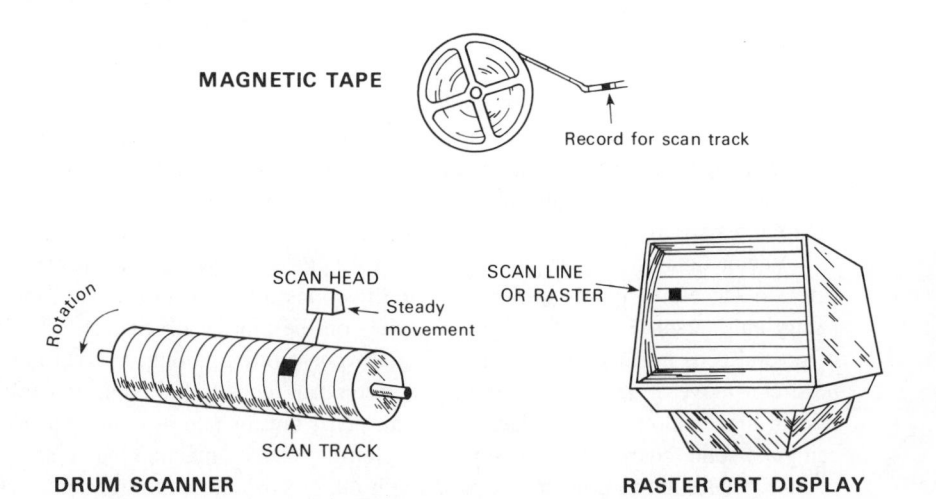

Figure 1-2 Sequential scan tracks of a map converted to digital data by a drum scanner (left) are stored as sequential records on a reel of magnetic tape (center) and displayed in sequence as scan lines or rasters on a raster-mode CRT unit (right).

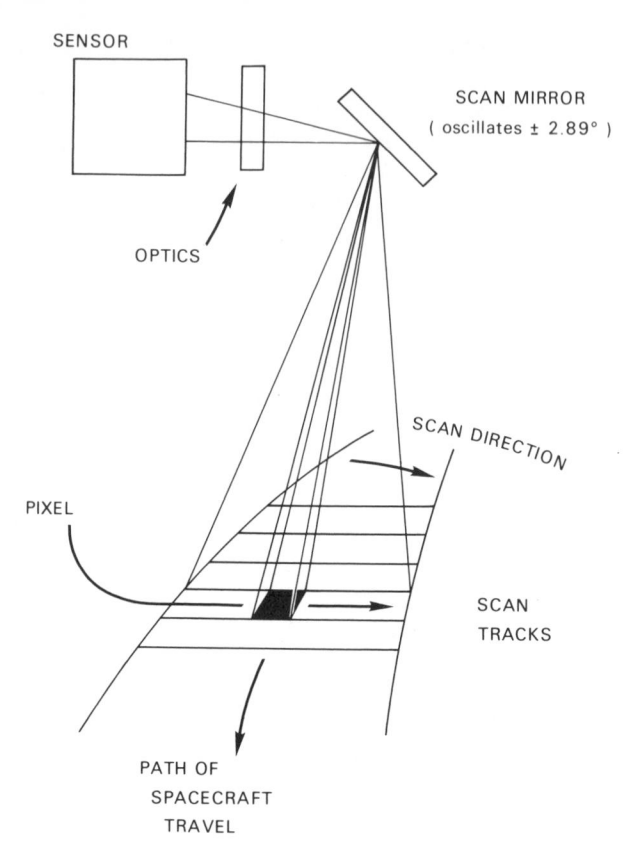

SENSOR

SCAN MIRROR
(oscillates ± 2.89°)

OPTICS

SCAN DIRECTION

PIXEL

SCAN
TRACKS

PATH OF
SPACECRAFT
TRAVEL

Figure 1-3 Scanner in an orbiting satellite is similar to a drum scanner: the oscillation of the scan mirror mimics the rotation of the drum by scanning all pixels in a scan track, and passage of the spacecraft over the ground resembles the movement of the scan head along the axis of the drum from one scan track to the next.

pixel or the number of pixels per unit length, with vector data the degree of detail depends upon the shortest line segment that might be displayed or recorded and the extent to which minute changes in direction can be represented. Because lines on maps have finite widths, the visual benefits of a highly precise display device might be apparent only if smooth, tight curves are to be plotted with narrow lines.

Vector data can be displayed readily by *flatbed plotters* that translate differences between successive points in the X and Y directions into synchronized movements of a gantry and cursor (Fig. 1-5). Electromagnets on the plot head can raise and lower one of several pens in fractions of a second, and when plotting a long, straight line, the plot head can move at speeds greater than a meter per second. Ballpoint and felt-tip pens sustain higher plotting rates than liquid-ink pens, but air pumps connected to ink pens can yield solid, sharp lines at comparable plotting rates by maintaining an adequate flow of ink. In addition to drawing on paper with ink or scribing (cutting) lines mechanically on opaque film, some flatbed plotters have a *photohead,* or *lighthead,* that exposes a line when the light is on during movement across photosensitive film. Dot-distribution maps can be plotted by moving to points with the light off and *flashing* a dot while the plothead is stationary.

Figure 1-4 *X-Y* coordinates are recorded with a digitizer by manually positioning the intersection of the crosshairs on the cursor over the intended point and pressing the record button. Numeric keys on the cursor promote the entry of category codes describing features as, for example, limited-access highways or unpaved roads.

Figure 1-5 Plothead routed through a coordinated series of displacements along the *X* and *Y* axes of a precision flatbed plotter can draw lines with a ballpoint pen or liquid-ink pen, scribe on scribing film, or expose lines on photographic film. A photohead with a lettering-symbol template can plot graphic-arts quality labels and point symbols on film. (Courtesy U.S. Geological Survey.)

The *drum plotter* is more common and convenient than the flatbed plotter because a continuous roll of paper on a rotating drum beneath a "fixed" gantry permits the plotting of several maps without a change of paper after each plot (Fig. 1-6). Supply and take-up reels coordinated with the rotating drum provide the paper tension required for rapid plotting. As with flatbed plotters, lines of different colors or widths can be plotted if the plothead has several pens. Vector data are important in computer-assisted cartography because pen plotters provide a relatively common and inexpensive means for producing publishable, fine-drawn maps.

Vector storage can be more economical than raster storage because only features relevant to the theme of the map or data base need be stored and processed. Raster storage is generally appropriate for a land-use data base that must account for all land within a region, whereas vector storage is likely to be more cost effective if, for example, only the paths of electric utility lines are to be recorded. In essence, vector data are more

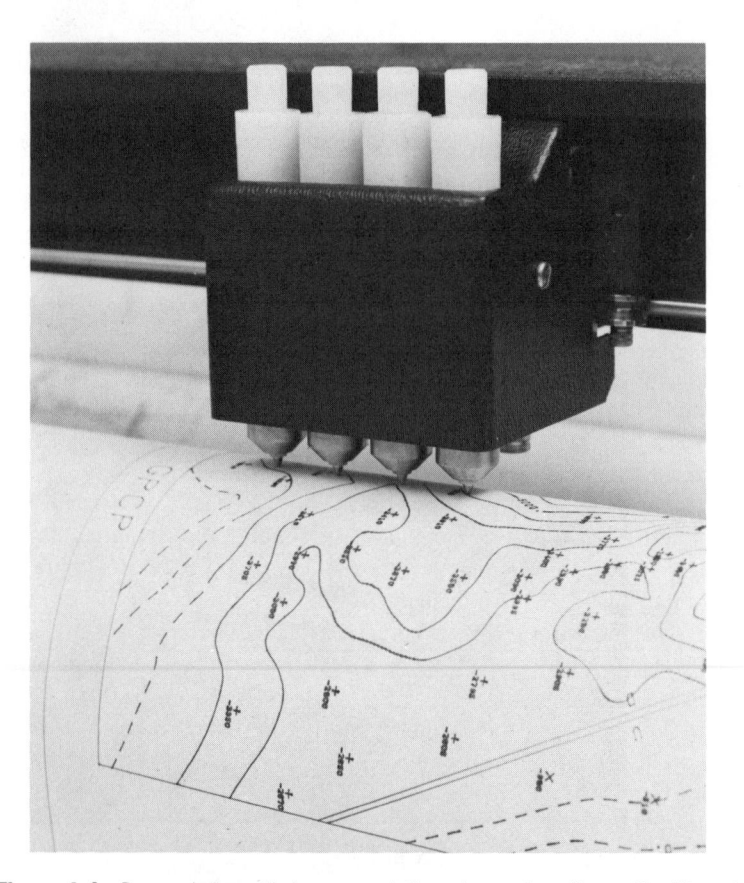

Figure 1-6 Drum plotter with four-pen plothead can draw lines with different colors or widths. Drum rotates beneath the plothead to provide movement in the *X* direction, and plothead shifts along track parallel to the drum for movement in the *Y* direction. (Courtesy California Computer Products.)

easily searched for features with particular feature codes, whereas raster data are searched more easily for all available information about particular places. Moreover, the economies of storing only point coordinates and feature codes diminish rapidly if the information to be mapped from a large geographic base file is more complex than a straightforward plot of the contents of the file. For example, if a rectangular map were to be prepared for a small, 120-km^2 area, each feature in a vector data base covering 120,000 km^2 might have to be examined. Features lying entirely within the region of interest could be plotted without any additional processing, but features lying only partly within the area would need to be searched further for their intersections with the boundaries of the region. With raster-mode storage this type of retrieval, called *windowing,* is much simpler.

Display devices generally are associated with a single type of data structure, but the cathode-ray tube (CRT) can display both raster and vector data. The CRT is particularly useful in the editing of vector files because of the rapidity with which maps can be displayed, altered, and displayed again for further inspection. Instead of the systematic scan sweeping 30 to 60 times a second from top to bottom on the *refresh tube* used for raster displays, the electron beam of a *storage tube* can be aimed freely anywhere on the screen. Image elements can be accumulated on the screen so that the cartographer can compile the content of a map, as well as preview its appearance (Fig. 1-7). A preview map might be thought of as *soft copy* to be inspected before using a pen plotter or other slower display device to produce a better-looking *hard copy.*

Refresh raster-mode CRT units are particularly suited for dynamic displays and have even been used to portray such animated maps as an oblique view of a three-dimensional surface rotating on the screen, with a continually changing viewing azimuth. Dynamic maps might also illustrate such spatial-temporal distributions as the advance of a settlement frontier or the variation in city traffic flows throughout the day. CRT displays generally are less suited to cartographic data than to alphanumeric displays, which do not call for fine resolution and a high rate of data transmission. Cartographic displays are also more demanding than typical applications in computer-assisted engineering design, where drawings usually are less complex and consist of fewer and longer line segments than required for most maps. For some applications, simplified geographic data bases are useful for efficient interactive mapping, whereas other applications require expensive, specially designed high-resolution display systems.

The microfilm recorder, sometimes called a laser-beam plotter or COM (computer output on microfilm) unit, is a natural extension of the CRT and is useful for the automated preparation of hard-copy maps. A laser beam can be used to expose fine lines on 35mm or wider film, as beam-forming optics control the diameter of the light spot, which can be positioned anywhere on the film plane by deflection mirrors or refracting prisms controlled by a small computer. The coherent light of a laser is less easily distorted in reflection and refraction than is white light, with many different wavelengths, and the sharp, high-resolution image can be enlarged greatly. Computer-controlled laser beams can plot lines at speeds greater than 5 m/s, and writing beams can be focused to diameters of 20 microns (10^{-6} m) or smaller. Press plates for offset printing can be produced from photographic enlargements of COM plots if the image size is small, or directly from the developed image if the recorded picture is sufficiently large. Laser-beam plotters have

Figure 1-7 Digital data entry stations with a digitizer and two CRT displays are used at the U.S. Geological Survey to record and edit "digital line graphs" based on existing printed maps and revision notes. Cursor on digitizer is also used to select edit functions from the menu, and electrostatic printer at left provides hard copy of CRT image. (Courtesy U.S. Geological Survey.)

been developed to plot maps that are then enlarged to 107 by 190 cm (42 by 75 in.), substantially larger than the standard topographic map.

Film recorders operating in raster mode can be used to prepare the color-separation negatives used in turn to make press plates for color maps. Consider, for example, a standard five-category choropleth map using colored area-shading symbols to portray differences among the states in the rate of population change. One category, colored blue, might include states that lost inhabitants, and four categories, represented by area shadings from pale to deep red, might illustrate increasing degrees of growth. Three press plates are needed: a blue plate for losses, a red plate for growth, and a black plate for state boundaries, title, legend, and neat line. Separate plots can be made for the base map and each color. For each category, fine dots can be plotted with a size and spacing appropriate to the intended shade of red or blue; increasing the dot size or decreasing the spacing of the dots will produce a brighter color. The dots are aligned in rows and are easily transferred to the film in raster mode. The images are kept separate until printed with different inks on the same sheet of paper. Secondary colors can be produced by printing two or more screened primary colors on the same part of the map sheet (Fig. 1-8). The dots might even be randomized to avoid the moiré effect likely if axes of the dot screens are aligned in approximately the same direction: if the plotting resolution is very fine, dots might be spotted at randomly selected grid cells.

Computer-controlled servomechanisms have harnessed COM plotting principles for data capture as well as for data display. In principle, a laser-beam vector plotter can be converted to a laser-beam line-following digitizer by replacing the unexposed film with a developed negative and installing a photomultiplier tube behind this negative. The light probe of this *flying-spot scanner* can thus follow a line by trial and error: when the light intensity sensed by the photomultiplier drops as the light probe strikes a darkened part of the film, the probe must retreat and attempt to trace the line in another direction. Only "successes" need be recorded, and the coordinates of points along a feature can be computed from the deflection angles of the mirrors aiming the light probe at the film plane. This inversion of data display for data capture is not unique to COM concepts: the electromechanical digitizer with a manually operated cursor sliding along a movable gantry to record X and Y coordinates is similar mechanically to the flatbed plotter. The drawbacks of mechanical inertia and equipment failure can be avoided, of course, by scanning the entire negative with a drum scanner and allowing software to track linear features.

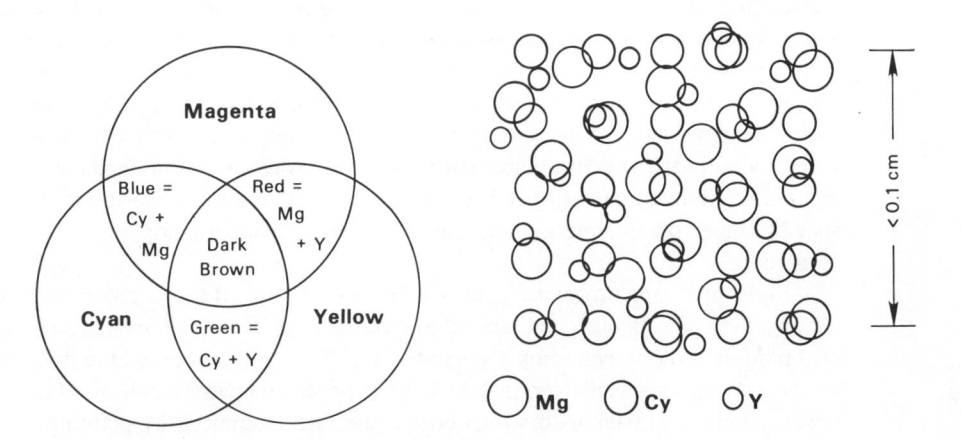

Figure 1-8 Subtractive primary colors (magenta, cyan, and yellow) can be blended to produce blues, greens, reds, and other colors (left). Fine-texture dot patterns of these primaries can be overprinted to produce a variety of blended hues, light and dark, brilliant and dull (right). When each pattern has 30 or more dots per cm, the eye responds to color, not to texture.

Once lines are plotted on a map, scanning digitizers examining only a limited portion of the drawing at any single time are neither totally reliable in following linear features across gaps and intersections nor consistently able to distinguish important features from unimportant ones. Geometrically irrelevant information, such as type, can be removed beforehand to provide "clean" map images for a scanner. Nonetheless, the assistance of a human operator is usually required to bridge interruptions, resolve uncertainties at intersections where directional trends are not reliable, and attach the codes needed to identify individual features. This human intervention might be provided in "real time," that is, while a vector-mode line-following digitizer is operating, or after digitization, in the case of a raster-mode scanner.

A CRT with a *light pen* is an effective display terminal for monitoring, editing, and annotating geographic base files. The light pen is a photosensor that detects the exact time at which the raster-scan electron beam of the CRT passes the location on the screen pointed to by the operator. The exact starting time and the rate of travel of the electron beam are known, and the elapsed time since the start of the scan can be converted to the plane coordinates of the indicated point. An operator with a light pen can thus tell the computer what parts of a figure to move, delete, connect, or relabel much more readily and accurately than if this information were entered manually through a keyboard. Positions on the screen of a CRT can also be referenced by a blinking dot positioned by a pair of thumbscrews, one for X and the other for Y. Some graphics systems position the reference dot with a *joystick,* a lever with two degrees of freedom (Fig. 1-1).

Tagging digitized features with identifying codes and adjusting the positions and linkages of digitized lines is promoted by edit stations with twin CRTs, one for display and the other for editing (Fig. 1-7). All labels and adjustments are tentative and are viewed first on the edit screen for approval before adding these changes to the data base. Voice decoders are sometimes used to receive verbal labels for map features indicated with a joystick or other pointing device. Digital elevation data recorded directly from photogrammetric plotters or orthophotoscopes obviate this later tagging of contour lines during an edit stage. With optically scanned contour maps, additional time might be saved by software which automatically assigns elevations to intermediate contours, so that only a limited number of index contours need be tagged, either manually or during an edit stage, by interpreting the elevation labels from existing topographic contour separations.

Light pens and joysticks can also be used to tell the computer what to do. A rectangular grid with each cell labeled to represent a different command can be displayed on a portion of the screen or on a separate CRT. Pointing to a cell within this grid, called a *menu,* invokes a specific command. A menu printed on paper can be used with a *digital tablet,* another digitizer with which coordinates are referenced by pointing to positions with a hand-held stylus. The tablet bed contains a printed-circuit grid of fine, closely spaced conductors. Electrical pulses sensed by a hand-held probe are unique for every grid intersection and are translated into *(x, y)* coordinates by the computer. Like a joystick, a digital tablet and stylus are sometimes used to reference locations on a CRT screen by controlling the position of a reference dot.

Linear data can be captured in vector format in either point mode or continuous mode. The operator of an electromechanical digitizer set for point mode positions the intersecting cross hairs of the cursor directly over each point and pushes a button or foot pedal to record the coordinates. Whereas point-mode data usually are screened to exclude redundant points, continuous-mode data frequently include numerous points that are either superfluous or represent minor manual errors in line following. When operated in continuous mode, digitizers frequently record points at constant intervals in distance or time. Usually many more points are recorded than are needed to describe the essential characteristics of a linear feature. Line generalization and smoothing programs that eliminate redundant and unrepresentative points can reduce substantially the computer time required for later processing.

The minimum distance interval for continuous digitizing and, indeed, the resolution of the digital tablet are specified by the grid of fine conductors beneath the platen. If the separation of the wires is 0.1 mm, nothing can be gained by attempting to record coordinates to the nearest 0.01 mm. The resolution can also be expressed as the number of addressable points per unit area. In this example there are 10,000 addressable points per square centimeter. The resolution of most display devices can be specified in this or a linear fashion, so that a laser-beam plotter with an overall plot size of 101 by 185 cm and 32 steps per millimeter would have nearly 2 billion independently referenceable points.

Matrix printers and ink-jet printers have a gray-level resolution in addition to a spatial resolution. A map is printed line by line in raster fashion on paper advanced past a row of sensitizers. Electrostatic matrix printers, which share some design principles with office copiers, have a row of small, closely spaced conducting nibs, or styli, that can charge the surface of a dielectrically coated paper. The resulting matrix of charged dots attracts a liquid or powdered toner in proportion to the electrical charge. An area with a relatively high charge will have larger dots and a darker image than an area with a lesser charge. Ink-jet printers produce a similar matrix image by spraying liquid ink row by row onto a moving sheet of paper. As many as 256 different gray-level patterns have been produced, many more than the eye can readily differentiate. The spatial resolution of some matrix printers can be equally impressive, rivaling the best halftone lithography of expensive magazines and presenting curved features with barely a trace of the raster-mode operation (Fig. 1-9).

If several colored inks can be used, the ink-jet process is particularly useful for the rapid display of hard-copy color maps. Colored toners can be used with the electrostatic process to produce results similar to color xerography. The visual efficiency of color mapping is available interactively, of course, with CRT units using principles developed successfully for color television.

Automated graphic display units can also mimic the dry transfer labels and point symbols used by cartographic artists and produce clear, sharp labels different from the crude, bland characters commonly produced by the pen plotter. With the advanced software *fonts* provided by some manufacturers of display hardware, the programmer need only specify type style, size, and position for letters, numbers, and special symbols such

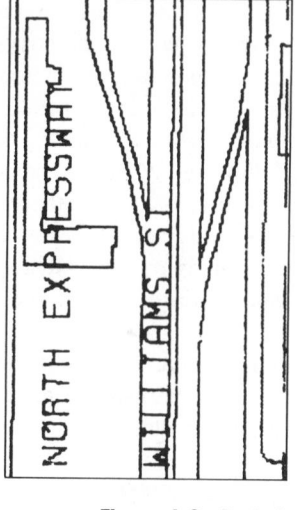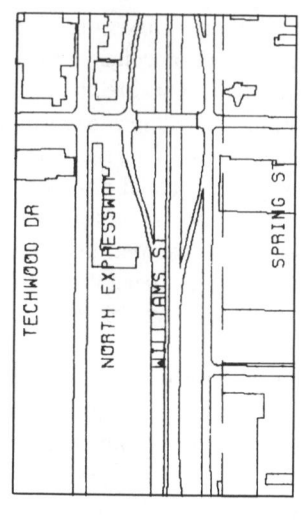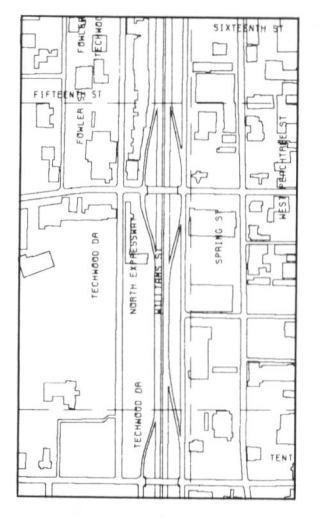

Figure 1-9 Part of an urban street and structures map plotted by a dot matrix printer with a resolution of 79 dots per cm (200 dots per inch). Center section is at scale of original plot; an enlargement to 200 percent, at left, provides a detailed view of the graphic marks, whereas the reduction to 50 percent, at right, illustrates the mollification by photoreduction of minor irregularities. (Courtesy Versatec.)

as punctuation marks. A variety of neat, crisp software fonts is available for high-resolution electrostatic matrix printers. Furthermore, small CRT photoheads can be mounted on flatbed plotters so that light and an automatically positioned lettering template can engrave type on photographically sensitive film, instead of drawing letters on paper with ink. Computer-drawn maps produced by state-of-the-art display hardware generally are not comparable in quality to the maps of the best pen-and-ink cartographers only because the controlling software does not yet mimic the trained mind of the draftsperson.

Computer-assisted cartography can still serve its two principal functions, geographic analysis and cartographic communication, without elaborate display hardware. Geographic analysis, for example, is assisted greatly by low-cost, flexible maps produced at an interactive CRT display, and a flatbed or drum plotter usually can produce efficiently whatever hard-copy displays the typical planning office might need (Fig. 1-10). The greater cost of higher graphic-arts resolution is not justified if the maps are to be either hung on walls for direct viewing or reproduced by inexpensive, monochrome offset lithography. Appropriate esthetic embellishments, if really necessary, can usually be added by hand. Expensive, high-resolution, state-of-the-art devices are likely to remain cost efficient only for the computer service bureaus, map printers, government mapping agencies, and industrial users able to distribute the purchase or rental costs widely over many maps. Even for these operations, more attention is likely to be given to the data bases and mapping software than to the display hardware.

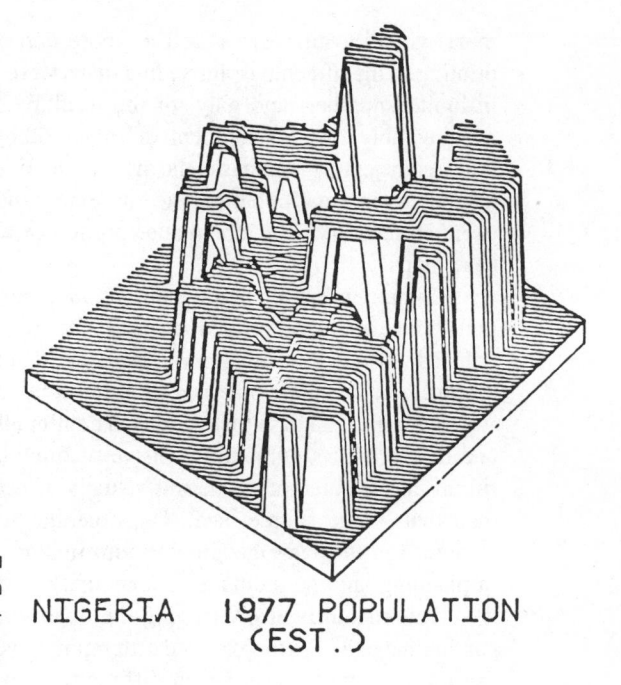

Figure 1-10 An oblique view of a population surface produced on an inexpensive flatbed plotter. (Courtesy Anthony V. Williams.)

NIGERIA 1977 POPULATION (EST.)

AN HISTORICAL PERSPECTIVE

Computer-assisted cartography, together with related developments in the remote sensing of land cover and land use from aircraft and satellites, is but one of a handful of technological innovations with a profound influence on mapping. Mapmaking can be traced to ancient Mesopotamia and the Nile Valley, yet people surely must have used available sticks, pebbles, bone fragments, and similar materials to illustrate geographic relationships even earlier than 2500 B.C., the established date of a regional map drawn on a clay tablet.[2] The first major innovation, in the second century, was Claudius Ptolemy's *Geography*, which described how the spherical Earth could be projected onto flat maps. Ptolemy's original maps did not survive, but his writings, when rediscovered after the Dark Ages, promoted a renewed interest in mapping during the fifteenth century. Also important as a stimulus was the European exploration of the Americas and the Orient. Navigational charting advanced notably after Gerhardus Mercator developed another significant innovation, the projection bearing his name, in the sixteenth century.

Several important innovations, of which automation is the most recent, made maps

---

[2]Norman J. W. Thrower, *Maps and Man: An Examination of Cartography in Relation to Culture and Civilization* (Englewood Cliffs, N.J.: Prentice-Hall, Inc., 1972), pp. 10–15.

more widely available, as well as more current and accurate. Before the invention of printing in the fifteenth century, fine maps were reproduced manually by copyists, slowly, in limited number, and only for the wealthy. Further advances in image reproduction, most notably the development of offset lithography and photography during the late nineteenth century, increased the availability of timely, accurate maps.[3] Without improved techniques in areal photographic surveying and photogrammetry developed in the early twentieth century, a much smaller portion of the world would have up-to-date, detailed base maps.

Computers are changing mapping in many ways. Soft-copy images, on CRT display units, have replaced printed maps for many applications. Production times have been reduced drastically, and for military intelligence a timely, high-resolution image of a designated site can be obtained in minutes by moving a nearby satellite into a new orbit. A geographic data base that is revised continually is available on demand for new editions and new products without expensive recompilation, and competent generalization algorithms and feature coding permit visually effective displays at a variety of scales smaller than that of the source data. The potential for linking files on both land and human resources in the same geographic information system is expanding the role of mapping in planning and the methods of map analysis in general.

Automation is also changing the quality as well as the amount of information that can be mapped. Textbooks on cartography have often stated that the cartographer should use no more than six or seven different area symbols when making a choropleth map. If the area symbols are cut from preprinted transparent sheets with a wax backing, a convenient technique in manual cartography, a limited number of categories is needed to assure visually distinct gray tones. This principle applies as well to line-printer maps, such as those shown in the next chapter to illustrate the CMAP program. Because an unlimited number of gray tones is impractical for this type of choropleth map, a classification is imposed so that each area symbol usually represents several different data values. Yet, when a computer and pen plotter are used to shade the areal units, a virtually limitless range of gray tones can be produced by varying the spacing of families of parallel lines (Fig. 1-11). Classification in choropleth mapping is no longer a necessity but merely an option available to the map author.

As might be expected, the new types of maps promoted by computer-assisted cartography have changed the directions of research on cartographic communication. For years, researchers have explored numerous methods of improving data classifications for choropleth maps. The advent of continuous-tone pen-plotter area symbols has demonstrated that the map user can receive more information than before about subtle geographic variations. Unfortunately, this information is likely to be distorted considerably during printing if relatively dark tones are plotted with thin shading lines. Moreover, in many situations the regionalization provided by a classification of data values can be of more

---

[3]Arthur H. Robinson, "Mapmaking and Map Printing: The Evolution of a Working Relationship," in David Woodward, ed., *Five Centuries of Map Printing,* (Chicago: University of Chicago Press, 1975), pp. 1–23.

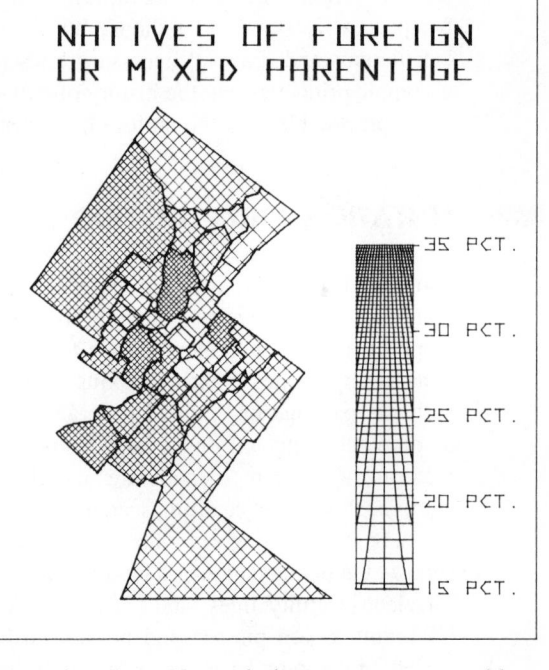

**NATIVES OF FOREIGN
OR MIXED PARENTAGE**

35 PCT.

30 PCT.

25 PCT.

20 PCT.

15 PCT.

Figure 1-11 Example of a no-class choropleth map with the tones of crossed-line area symbols proportional to foreign stock percentages of the populations of census tracts in Scranton, Pennsylvania, for 1970. (Source: Mark S. Monmonier, "The Significance and Symbolization of Trend Direction," *The Canadian Cartographer,* 15, no. 1 [June 1978], 35–49. Reprinted by permission of the publisher.)

use than a map revealing minor differences in values. New solutions present new problems and need not obviate old solutions.

Continuous-tone shading is not new. A cartographic textbook published in 1949, well before the crudest computer-produced map, includes several examples of "proportional shading" with the spacing of horizontal lines inside areal units inversely proportional to data values.[4] These ideas may even be traced back to 1832 and the French cartographer Adolphe Quetelet.[5] Continuous-tone shading was little used until 1973, when Waldo Tobler proposed cross-line patterns with proportional graytones as more appropriate for choropleth maps than a small number of graytones portraying discrete categories.[6] How many other significant rediscoveries await the cartographer who appreciates both the potential of the computer and the history of mapmaking?

Automation also has made cartographers less reluctant to choose map projections more suited to a particular application than those of readily available printed base maps. Drafting by hand the meridians, parallels, coastlines, and political boundaries of a pro-

[4]T. W. Birch, *Maps: Topographical and Statistical,* 1st ed. (Oxford: Oxford University Press, 1949), pp. 166–67, 186–87.

[5]Alan M. MacEachren, "The Evolution of Thematic Cartography: A Research Methodology and Historical Review," *Canadian Cartographer,* 16, no. 1 (June 1979), 17–33.

[6]W. R. Tobler, "Choropleth Maps Without Class Intervals?" *Geographical Analysis,* 5, no. 3 (July 1973), 262–65.

jection tailored to minimize distortion in important areas or provide a particularly intriguing perspective is slow and tedious, but a computer and plotter can perform all needed calculations and draw all necessary lines in minutes, not days. An appreciation of cartographic principles and the history of cartography is important because the computer has made practicable what heretofore had been merely possible.

IMPLEMENTATION AND PLANNING

Among other changes, the computer has accentuated the need in cartography for careful planning. Difficulties in planning for automated mapping relate more to the type of mapping system implemented than to the specific geographic application. At least three general approaches exist to making computer-assisted cartography operational: (1) the software package or "canned program" usually made available in a program library at a computer center serving a large number of users; (2) the turnkey system, a complete, ready-to-run system of software and hardware, possibly including highly specialized equipment such as a drum scanner or an orthophotoscope, and designed to be used exclusively by a single private firm or government agency for a specific mapping task, such as the maintenance of a land-resource data bank or a geographic information system for electric utility lines; and (3) a small multipurpose system, with a minicomputer, a CRT unit, a pen plotter, and possibly a digitizer, to be used by several persons for a variety of mapping tasks, some known, others not yet known when the system is installed.

All three approaches can lead to dissatisfaction with the maps that are produced and the maps that are wanted but cannot be produced. Technological change, such as the development by another manufacturer or software vendor of a better, more versatile product at a lower cost can aggravate the user wanting the latest, best, and most versatile system, but a more serious problem with obsolescence is an unanticipated, increased demand for mapping services in the classic sense of a service or product becoming "too successful" and requiring considerable additional capital investment. Many difficulties, however cloudy the future might appear at the time decisions are made, can be avoided through a careful analysis of organizational and client needs, hardware capabilities and trends, cartographic principles, and current software development.

Mapping software packages can be designed, programmed, and debugged by and for an individual user or group of users or obtained from another installation. The "imported" mapping package might be received free of charge from a friend or acquaintance; provided at only a nominal cost, say, under $50 to cover magnetic program tape, postage, and handling, by a government agency; or purchased at a moderate or substantial cost from a commercial developer of graphic software or a university-related cartographic laboratory. (A list of sources for mapping software and other information related to computer-assisted cartography is provided in the appendix.) In the latter case, the purchaser might acquire from the commercial vendor only the right to use the program for a stated period of time, with additional payments required to extend the lease.

United States copyright and patent legislation and case law have not afforded the software developer the protection granted the novelist and poet, and the customer often is required to sign a statement pledging that the package will not be used on another computer or copied for another user.[7] The client might also be required to accept a disclaimer absolving the vendor from any claim for damages should the program produce erroneous results, upon which erroneous decisions might be based. As with any legally binding agreement, the purchaser of an expensive mapping program would do well to have an attorney examine the provisions of the sales contract.

Commercial software vendors frequently agree to provide, at cost, additional copies of users' guides and other important documentation, to notify the customer of any errors detected in the code or algorithm, and to supply, free or at a nominal charge, all updated, modified, or expanded versions of the program. Assistance might also be provided to users attempting to adapt the program to a computer system other than the model with which the package was designed and tested. At a minimum, a sample set of representative data and corresponding results should be provided for testing by the new user. Indeed, these support services alone often justify what might seem a relatively high price for a tape copy of a program.

User documentation and training aids are important, particularly if the program is likely to be used by persons unfamiliar with computers or mapping. Instructions should be clear, concise, and not ambiguous. Where appropriate, examples should be included. Users with different academic backgrounds, such as architecture, civil engineering, geology, marketing, or planning, might benefit from manuals written from a disciplinary perspective.

Unless the user specifically selects the approach to a mapping problem, a canned program commonly will follow one particular method—by default—even though the software might be useful for a much more general range of problems. Users not familiar with the basic principles of cartography ought never to be encouraged to trust blindly these *default options*. A cartographer or programmer hundreds of miles away, with an inherently limited perception of the possible future uses of the package, cannot be expected to anticipate the needs of all users, except by providing the neophyte a flexible program with clear instructions and all appropriate caveats. Mistakes are quite likely if you allow someone who knows nothing about your data and objectives to design your maps.

Mapping programs can be developed for a wide variety of cartographic operations, but no software package is likely ever to include all possible mapping methods, or even a large but representative variety of mapping techniques. Mapping software packages commonly produce one type, or a few related types, of map, usually for a single class of display device such as the CRT unit or the line printer (Fig. 1-12). Brassel, in a survey

---

[7]Roy G. Saltman, *Copyright in Computer-Readable Works: Policy Impacts of Technological Change* (Washington, D.C.: U.S. Department of Commerce, National Bureau of Standards Special Publication 500-17, 1977).

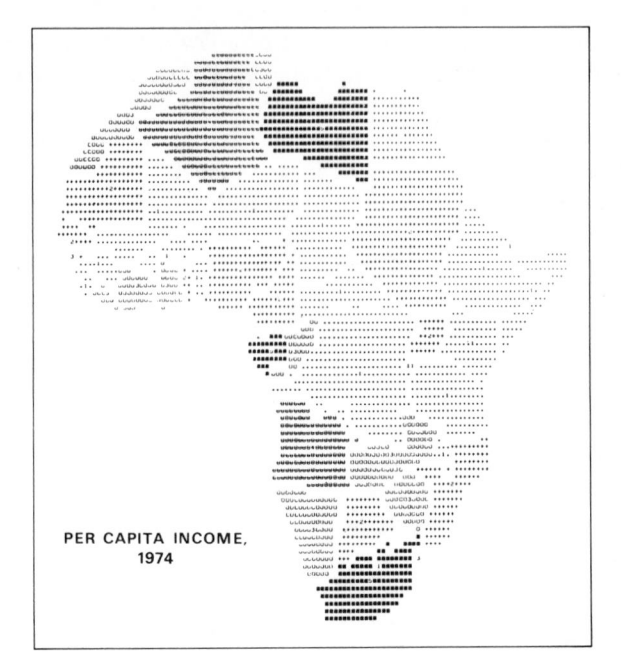

PER CAPITA INCOME,
1974

Figure 1-12 SYMAP, a widely available mapping package, produced this income map on a line printer, a standard part of most large, general-purpose computing systems. (Courtesy Anthony V. Williams.)

of over 200 mapping packages, recognized 11 distinct groups of programs.[8] His categories are as follows:

1. Data collection, editing, and manipulation.

2. Basic drafting operations.

3. Diagram display.

4. Display of point and line features.

5. Area shading programs, subdivided according to raster mode and vector mode.

6. Contouring programs.

7. Three-dimensional representations.

8. Map projections, transformations, and distance measures.

9. Miscellaneous cartograms.

10. Display systems.

11. Geographic information systems with graphic display capabilities.

---

[8]Kurt Brassel, "A Survey of Cartographic Display Software," *International Yearbook of Cartography,* 17 (1977), 60–77.

For many prospective users, mapping software is second in importance to mappable data.[9] Fortunately for some users, all or part of the needed data can be purchased as easily as the programs to process it. Two distinct types of data are often needed: (1) spatial data files describing point locations, area boundaries, and linear features, and (2) statistical or thematic data files providing qualitative or quantitative descriptions of places. For many applications the thematic data will change completely or require substantial updating, whereas the spatial data will be static or require very minor alterations. The prudent user will coordinate purchases of spatial data files and mapping software to ensure compatibility. Some software vendors provide spatial data already organized for direct use by their programs, or in some cases provide conversion programs with which the user can restructure the data. For complex statistical files, such as those reporting areally aggregated results of the Census of Population and Housing, retrieval and analysis packages are available for developing refined files of thematic data for mapping.

Many useful data sets are in the public domain and can be obtained at low cost from government agencies. In addition to the statistical files for its economic and population censuses, the U.S. Bureau of the Census can provide DIME (Dual Independent Map Encoding) files describing the street network, blocks and tracts, and addressing schemes of larger urban areas. Census-tract outlines for major metropolitan regions are available in the Bureau's Urban Atlas Files. The Bureau also is the source for magnetic tape copies of its *County and City Data Book,* a standard reference for geographic statistics. World Data Banks I and II, developed by the Central Intelligence Agency to cover the entire globe, describe coastlines and international boundaries. Files developed in the Department of Transportation describe county boundaries for the entire United States. Among the cartographic data bases maintained by the U.S. Geological Survey is a digitized map of the entire country, which can be displayed at scales of 1:2,000,000 or smaller, and which includes county boundaries, population centers, highways, railroads, and drainage lines. The National Cartographic and Geographic Information Service (U.S. Geological Survey, Reston, VA 22092) is a clearinghouse for all requests for information about federal mapping activities and provides digital map data as well.

Some applications, such as regional land-use inventories, urban cadastres (records of real estate ownership and property assessments), and inventories of highway accidents, are sufficiently specialized to require mapping and analysis systems designed to serve specific narrowly defined needs. The staff of the organization managing a geographic data system, with the advice of consultants, might design an appropriate configuration of hardware as well as write the necessary software. Occasionally, a research and development firm is awarded a contract to provide a fully operational system. These highly specific, *dedicated* graphics systems commonly are oriented primarily toward data capture, with digitizing and editing as the principal uses. Analytical operations requiring a larger

---

[9]Eric Teicholz and Julius Dorfman, *Computer Cartography: World-wide Technology and Markets* (Newton, Mass.: International Technology Marketing, 1976).

internal memory might be carried out with a large batch-processing computer or possibly a time-sharing system in another part of the city linked by telecommunications wires to the smaller, interactive data management and display system.

Dissatisfaction with a dedicated system might be more deeply rooted than complaints about the efficiency of the hardware or the versatility of the software. Machinery can be replaced or expanded and programs can be rewritten or modified, but structural weaknesses in large data files, developed at considerable time and expense, are not so easily rehabilitated. Fundamental flaws of geographic data bases include an overly coarse spatial resolution and inordinately general thematic descriptions. As examples, a land-use information system with 1-km^2 grid cells is too imprecise for analyzing metropolitan areas, and in most places a local environmental resources system defining wetlands solely on the basis of elevation would either not describe adequately the habitats of endangered hydrophytes or include land generally suitable for development. Planning a map analysis system involves far more than ordering a computer and its peripherals and writing the programs: a system must be designed to serve a clearly stated objective, and preliminary investigation often is needed to define the cartographic requirements of what frequently is a noble but vague goal.

Microcircuitry and the economics of mass production have lowered the cost of an efficient microprocessor into the price range of the expensive toy. Although complete, working systems can be purchased from a single vendor, a comparable hardware configuration might be obtained at a much lower cost by purchasing the components separately from various manufacturers or electronic equipment retailers. Despite the higher initial cost, the complete system, purchased with guaranteed interfaces among the various components, might effect another type of savings through reduced frustration and downtime.

An interactive system usually can be expanded later, but the cost of adding a new display unit or additional memory might even be more than that for an entire system superior to the one modified. Rapid technological development has increased the likelihood that products will become obsolete in a few years, discontinued, and no longer available for upgrading a system. Some components, of course, are more likely to be rendered obsolete than others: a small flatbed plotter with decent resolution can be used for many years, but a central processor able to keep pace with such an electromechanical plotter might be noticeably sluggish if used to drive a CRT display. The plotter driven by a tape reader, with the tape generated on an independent computer, might afford greater flexibility insofar as plotter tapes can be prepared on different computers. The conscientious buyer will carefully consider present and future needs for internal memory, processing speed, plotting and digitizing large drawings, multiple external memories, alternative arrangements in the event of hardware failure, time sharing, hard-copy and CRT displays, and light-pen editing, to name but a few details.

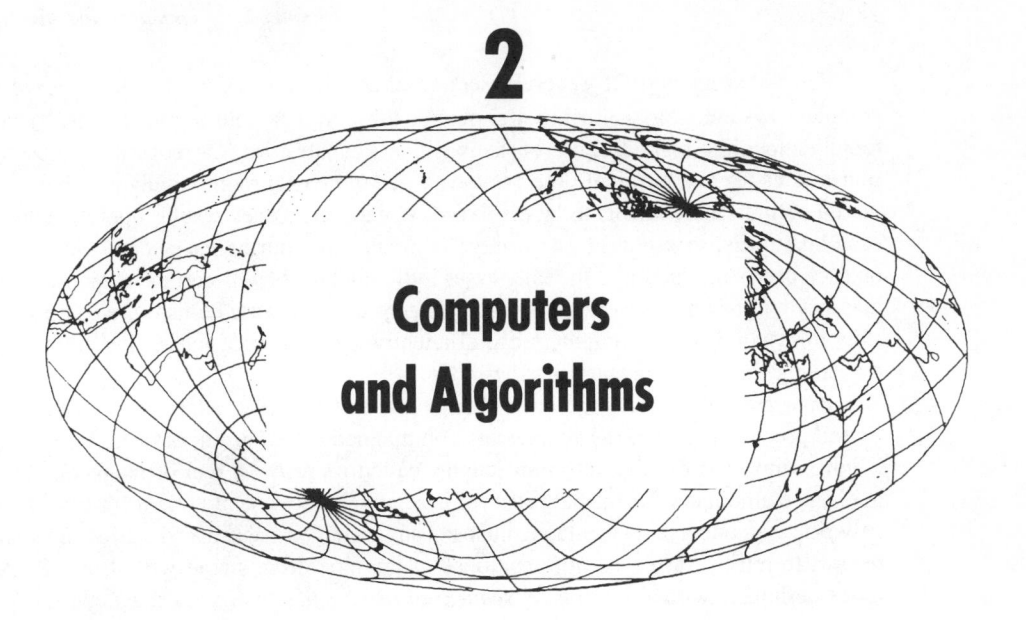

2

Computers and Algorithms

An appreciation of the potential of computer-assisted cartography requires an understanding of elementary computer architecture and program structure. This chapter introduces some basic terms and principles related to computer memory and logic, and provides a theoretical basis for structuring many mapping procedures as programmable conceptualizations called algorithms. The use of algorithms and programming is illustrated with CMAP, a simple FORTRAN program based upon the highly useful scan-line concept. CMAP produces crude maps on the line printer, but the concept is useful in color graphics.

MEMORY AND LOGIC

Devices for data entry and display, called *peripherals,* are subordinate to the computer's control unit, or *central processor.* The central processing unit (CPU) has two principal components, an arithmetic logic unit (ALU) that performs all arithmetic and logical operations and a control unit that monitors the transfer of data among the various components of the system. A basic computer system consists of a central processing unit linking a memory and one or more peripherals for data entry and display. A peripheral unit with both a keyboard and a printer or CRT, for two-way communication with the computer, is called a *terminal.* Remote terminals linked to a computer through telecommunications lines can be used to direct computers in distant cities. Some terminals, the "smart" terminals, also have a small memory and rudimentary CPU.

Time-sharing permits several users to share the resources of a large, expensive computer system. Persons using a computer through a remote terminal cannot respond rapidly enough to make full use of the machine's capabilities. Moreover, a computer can transmit results to a terminal, line printer, or pen plotter far more rapidly than mechanical inertia permits these results to be displayed. Greater efficiency results when several users each take turns, in sequence, on a large computer that during each round works briefly on each user's instructions. In many cases individual users will find the response so rapid that they will be unaware of others sharing the system. Because humans are comparatively slow to respond, large computers can efficiently accommodate interactive graphics only by serving the user in a time-sharing mode.

Data are available to the computer in *internal memory,* addressed directly by the central processor, or in *external memory,* on magnetic tape, drum, or disk. Tape memory is long, thin, and divided into unit lengths called *records.* A reel of magnetic tape can hold large amounts of data, but retrieval is relatively slow. Only one record can be read or written at once, and the tape must be advanced or backspaced across intervening records to retrieve a record not positioned at the tape drive's read-write head. Retrieval is less efficient with this type of *sequential-access file* than with magnetic disks and drums, which may be organized as *random-access files.* A drum rotates rapidly about its axis and data can be read from its magnetized surface in the small fraction of a second taken to bring the desired portion of the cylinder under the long, linear read-write head parallel to the axis of the drum. Access time is not appreciably longer with a disk drive, in which a mechanical arm containing a read-write head must be positioned above the appropriate track on the rapidly spinning, magnetized disk.

Although access can be slow, tapes are inexpensive, easily mounted or removed by the operator, and provide the most common method for exchanging programs and data. Drums are removed only for maintenance, and disk packs containing several rigid disks joined to a common axis can be mounted or dismounted readily, but are too expensive for the average user. Relatively inexpensive *floppy disks* are manufactured of flexible plastic with a magnetic coating and provide convenient random-access storage for individual users with limited resources. Data and program instructions can also be stored on cassette tapes and punched cards, with numbers and letters coded as a series of holes punched in vertical columns.

Nonmechanical additions to a computer's main memory are possible in the form of *bulk storage units,* supplementary memory devices with rapid electronic storage and retrieval because there are no moving parts. For still more rapid transfer of data from a memory unit to a peripheral, *direct-memory access* momentarily transfers control of the computer to a peripheral such as a raster CRT unit, which then receives data directly from the memory unit. Time is saved by not routing large quantities of data through the CPU control unit. A group of related data entries can be organized as a *file* and referenced as a unit with the file name.

The integrated circuit has blurred the distinction between hardware and software. A circuit describing either a computable function or some data can be photographically reduced and etched onto a silicon *chip,* a tiny fraction of a square centimeter in area. A chip is comparatively inexpensive to produce and performs the functions of thousands of transistors. Data and programs thus are both available in material form as *firmware*, with features of both software (coded information) and hardware (physical equipment). *Read-only memory* units (ROMS) containing programs or data that cannot be erased can be added to a system when needed. Users can develop their own firmware with programmable (PROMS) and erasable (EROMS) read-only memory units. Large-scale-integration (LSI) technology, which permits the production of even such complex circuits as a central processor unit as a single integrated circuit, has greatly reduced both the cost of computers and the floorspace required for a sophisticated, powerful data-processing system. An efficient microprocessor might require no more space than half a desktop.

The basic unit of information used by a computer is the *bit,* an abbreviation for binary digit. Each bit has a value of either 0 or 1, and a string of bits represents an integer by starting with a one's place and proceeding to the left with a two's place, a four's place, an eight's place, and so forth. In binary arithmetic, 0001 represents 1, 0010 represents 2, and 1010, with digits in the eight's and two's places, represents 10. Binary coding represents numbers in computer memory by a series of presences or absences of electronic charges. Because many early computers stored data by setting the magnetic polarization of small ferrite rings called *cores,* internal memory is often called *core storage.*

Retrieving stored information, even from internal memory, takes time, and data storage and retrieval of individual bits is inefficient. In most computers the smallest unit of memory that may be referenced is the *byte,* usually a string of eight bits. Several bytes can be joined together to represent a single, larger addressable unit of memory called the *word.* Integers usually are represented by two- or four-byte words, although some computers permit other word sizes. The binary value of a standard two-byte word, with the first bit of the first byte reserved for the sign, cannot exceed $\pm 1111111,11111111$, that is, an absolute value of 32,767 or $2^{15} - 1$. If decimal places are needed, a *floating point* number, with one byte reserved for the exponent of 10, is used. For example, -17.235 can be treated as -0.17235×10^2 and represented in binary form by a mantissa of $-0000000,01000011,01010011$ and an exponent of $+0000010$. The maximum absolute value of the three-byte mantissa, with one bit allocated to the sign, is $2^{23} - 1$, or 8,388,607, and the maximum absolute value of the single-byte exponent is $2^7 - 1$, or 127. Thus the largest number that might be represented in this fashion is $\pm 0.8388607 \times 10^{127}$, and the smallest nonzero number is $\pm 0.0000001 \times 10^{-127}$. A *double-precision* word with seven bytes allocated to the mantissa would widen this range to between $\pm 0.36028797018963967 \times 10^{127}$ and $\pm 0.0000000000000001 \times 10^{-127}$. These limits are not uniform, and some computer manufacturers have set more restrictive limits.

Differences in word size explain why the same program run on different computers with the same data might produce different results. Consider a simple sequence of arithmetic operations: subtract two numbers, divide by a third number, and then subtract this result from a fourth number; for example,

$$\left(\frac{11,112 - 0.123}{0.66} \right) - 16,836.17$$

The result of the first subtraction would be truncated to 11,111.87 by a computer with a 32-bit word, but would remain 11,111.877 for a machine with a 36-bit word. Subsequent division by the denominator will yield truncated results of 16,836.16 and 16,836.177, respectively. The final subtractions in this admittedly contrived example would produce greatly different results: -0.01 for the 32-bit machine and 0.007 for the computer with the larger word length. Note that, however precise the second result might appear, the true answer is an irrational number, 0.007272727 . . . , which would require an infinite number of bits for its exact representation.

Because many programs, particularly for statistical analysis, involve long chains of calculations, the word size specified by the manufacturer might produce grossly misleading results if the precision—double or single—specified by the programmer is not appropriate for the operation. Skillful programming can, of course, avoid many problems with numerical precision. For example, the algebraic expression

$$\frac{\left(\dfrac{(XY)}{X}\right) - W}{Z}$$

can be reduced to

$$\frac{Y - W}{Z}$$

before coding the program, thereby avoiding the rounding error of the previous example. Yet the expression is more appropriately coded as

$$\frac{Y}{Z} - \frac{W}{Z}$$

if Y is much larger than W. If Z is greater than 1, division by the denominator before subtraction can defer the rounding that otherwise might approximate $(Y - W)$ by Y alone.

Other difficulties related to word size are *overflow*, when a positive exponent exceeds the allowable limit, and *underflow*, when a negative exponent falls below the lower limit. These errors, which usually halt execution and yield an appropriate *error message* to the user, might be avoided by using double-precision specifications for all numbers. Yet the storage then required to execute the program might exceed that available, whereas a well-designed program would accommodate a typical range of data values without the indiscriminate, wasteful use of internal memory.

Both program and data consume memory, and it frequently is necessary to ask whether a computer is large enough to accommodate a particular mapping program. The size of a computer's internal memory is commonly measured in bytes or, more specifically, in kilobyte units. A kilobyte is not 1,000 bytes, but 1,024 bytes, 1,024 being 2^{10}.

Computers can reference storage locations and assign memory in binary-based chunks because the addresses of particular words in memory are stored or computed as binary numbers. The symbol K, as in 48K, is often used to indicate kilobytes, but sometimes refers to kilowords for computers not designed to allocate memory in bytes. A computer with less than 4K or so of memory and a word size of 18 bits or less is called a *minicomputer*.

It is convenient to think of the CPU as a device that extracts data values from memory, alters them or creates new ones, and stores the results. The CPU must be told where in memory to go for the next operation, as well as where to go for the next data value. Thus some memory locations are used to specify the bit-encoded machine instructions and others point to the location in memory of the next instruction. This coded recipe for calculating results is called a *program*. Like recipes for preparing food, a program consists of a series of these discrete steps and may include commands to accept new data, display results, and jump to another part of the program if certain conditions are met. For many persons the most difficult aspect of learning to write computer programs is decomposing a process into this necessary sequence of discrete steps.

Some programs are represented diagrammatically by *flow charts* consisting of a linked series of boxes with the directions of flow indicated by arrows. These boxes represent operations to be performed by the computer, and the flow lines indicate the sequence in which the operations are to be performed. Branch instructions might be used to compare the values of two numbers and determine which of two possible directions the computer is to follow next. Branching tests are called *logical operations,* as distinguished from arithmetic operations, because the condition tested is stated as an equality or inequality, which can be either true or false. If a comparison such as "X greater than Y" ($X > Y$) is true, the computer might, for example, be directed to subtract Y from X; otherwise, when the condition is false because Y either equals or exceeds X, some other action would be indicated. Flow charts represent branch instructions by diamond-shaped boxes with two alternative departing directions labeled "yes" and "no."

For convenience, flow charts and computer programs often treat related numbers, such as a list of numbers to be added, as members of an array. Each array has a single name, for example, X, and individual positions in the list are referenced as X(1), X(2), X(3). . . . A variable subscript or array index such as J can also be used so that X(J) references the tenth position in array X when J has the value of 10. Computers store two types of number: *variables,* which are represented by an alphanumeric name such as X, J, POP60, and LISA, and *constants,* which are merely values specified directly in the program. A variable may be a single number or an array of numbers with individual elements referenced by subscripts as shown. Arrays may have more than one subscript; for example, if A is a two-dimensional array, A(10,3) refers to the number stored in the tenth row and third column of A. Subscripts are always integers.

The flow chart in Figure 2-1 illustrates how a computer adds up the numbers in an array by accumulating a running sum. A single-value, that is, unsubscripted, variable SUM is used to accumulate the total. SUM must be set to zero in an operation akin to clearing the register of a simple pocket calculator prior to adding several numbers. The

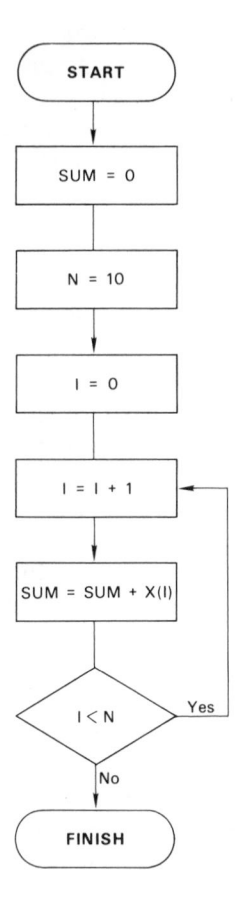

Figure 2-1 Flow chart for accumulating sum of ten numbers in array X.

numbers to be added in this example are the ten elements of single-dimension array X, assumed here to have been stored in memory by a previous program. *Replacement operations,* rectangular boxes with expressions containing equal signs, are used to set to 10 the variable N indicating the effective length of array X and to 0 the index I used as a counter for array X.

After these *initializing* instructions, the flow chart proceeds to a *loop.* Each pass through this loop adds another number stored in X to the running total in SUM. The index I is *incremented* at the beginning of each pass by adding 1 to the current value of I and then storing this result as the new value in memory location I. In the next step the values of variable SUM and the Ith position in array X are retrieved from memory and added; this new result for variable SUM is likewise used to replace the previous value. The values of I and N are then compared, and another pass through the loop is indicated as long as the condition I < N is satisfied.

COMPUTERS AND GENERABLE BEHAVIOR

Basic to developments in computer-assisted cartography is the principle that, if an operation can be decomposed into a finite number of simple steps, the operation can be carried out by a computer-controlled machine. The concepts introduced in this section provide a theoretical basis for the development of highly sophisticated mapping programs capable of surpassing in many ways the skills of the sometimes inattentive, easily bored, frequently inconsistent human cartographer. Both the human and the computer may be conceptualized as finite-state machines, the human having more states and often behaving creatively, the digital computer having fewer states and being more easily harnessed to specific goals.

The concept of a *finite-state machine* can be demonstrated through a simple simulation of the behavior of a computer. Take a sheet of paper and draw a small box labeled N and two large boxes labeled I and SUM. These boxes represent addressable locations, or words, in computer memory. Also represent the array X, in this case by 10 small boxes stacked in a vertical column. Write numbers in these 10 boxes, any numbers you like. Now proceed to the first step in the flow chart (Fig. 2-1) and write a zero in the box labeled SUM. Write small figures because each time the flow chart tells you to store a new value in SUM, or in I, the previous value must be crossed out.

Follow the steps outlined in the flow chart. Follow these steps rigorously, as a computer—a dumb but fast robot—must follow them. On your paper write a 10 in the box for N and a small 0 in the box for I. Then begin the loop by replacing the 0 in I with a 1. Next extract 0 as the current value of SUM, add it to the value stored in X(1), and write the result in the SUM box after crossing out the 0. Your first test condition, Is 1 less than 10?, is satisfied, so repeat the loop. Follow the flow chart blindly; it should lead you through the loop nine more times, after which the box labeled SUM should contain the total of the ten numbers in array X.

The computer memory, as represented on the sheet of paper, passed through several stages during the simulation. Each time a new number was written the computer moved to a new state, as indicated in Table 2-1. This *transition table* is an abbreviated representation of the various states encountered in adding the even integers between 2 and 20. The behavior of any machine with a finite number of states can be represented by a transition table.

The British mathematician A. M. Turing advanced the idea that the human mind, which has only a finite number of states, is a finite-state machine and thus similar to a *digital computer*. (The computers discussed in this book are digital computers, essentially counting machines, as distinct from *analog computers* in which quantities are represented by voltages and are thus subject to significant drift and other distortions caused by the temperature and humidity of the computer's immediate environment.) Turing also developed the concept of the *Turing machine,* a control unit with a finite number of states and a read-write head for storing and retrieving information with a tape of infinite length (Fig. 2-2). The transition table of a Turing machine uses the current internal state of the memory and the information read from the tape record directly below the read-write head to determine not only the next state of internal memory but also an optional replacement

TABLE 2-1   *A simple transition table for adding even integers between 2 and 20*

State	State variable		
	N	I	SUM
1	—	—	0
2	10	—	0
3	10	0	0
4	10	1	0
5	10	1	2
6	10	2	2
7	10	2	6
.	.	.	.
.	.	.	.
.	.	.	.
20	10	9	72
21	10	9	90
22	10	10	90
23	10	10	110

value for the tape record and the number of records the tape is to be advanced or backspaced before the next transition. This elaborate transition table is called an *algorithm*.

Turing's theories hold that an appropriate machine and algorithm can be constructed to perform any process than can be decomposed into a finite number of discrete steps. Included are operations that might generally be called "thought," although not necessarily original or creative. The term algorithm refers broadly to any set of instructions that might be followed without original thought by a machine or person. The similarities in switching behavior of the digital computer and the human nervous system suggest further that machines can be programmed to respond to a variety of complex situations in ways similar to the reactions of humans. In fact, psychologists have written programs to simulate the responses of both normal and paranoid personalities to queries typed at remote terminals. Until thinking is defined precisely, it is difficult to state categorically that computers cannot be made to think.

This concept of *generable behavior* can be extended to the design of maps, and programs might even be written to mimic the ideal responses to various mapping problems of different schools of cartographic thought. Automation thus might revolutionize cartography by concentrating research efforts on problems in map communication likely to yield results easily incorporated into mapping algorithms. Decision-making algorithms responsive to a wide variety of needs and constraints can be appended to efficient, widely distributed mapping software, thereby giving many map makers and map users access to the most current cartographic wisdom. The cartographic practitioner might become very like the dispensing physician, diagnosing a client's needs and then prescribing an appropriate and easily obtained solution. In this analogy the cartographic researcher would play the roles of biochemist and research physician, performing basic research on the mechanisms of map communication and analysis. As digital computing becomes more common, as home computers become larger and more efficient, and as peripheral display devices become less costly and more available, there might be little need for anyone but the artist to lift a drawing pen. And even the artist will appreciate the opportunity to call a drawing from a computer's memory, add a few important finishing touches, and obtain a finished map in minutes.

This picture of the map business assisted fully by computers is neither unrealistically futuristic nor inherently diabolical. Most maps are highly stylized art, quite amenable to automated display. Although computer algorithms can simulate the trial-and-error thought and evaluation of the conscientious cartographer, imagination has a definite place in this scenario, a place that affords creative ideas a rigorous testing and wide dissemination. The cartographic artist using the pen and ink of folk cartography would prepare few mundane maps for statistical reports, preferring instead the satisfaction of carefully conceived, highly individualistic map representations of the physical and human landscapes. The prospect of widespread map abuse is also present, and control through professional licensing and regulatory bodies would appear likely. Some changes will be good, others bad, but change is inevitable.

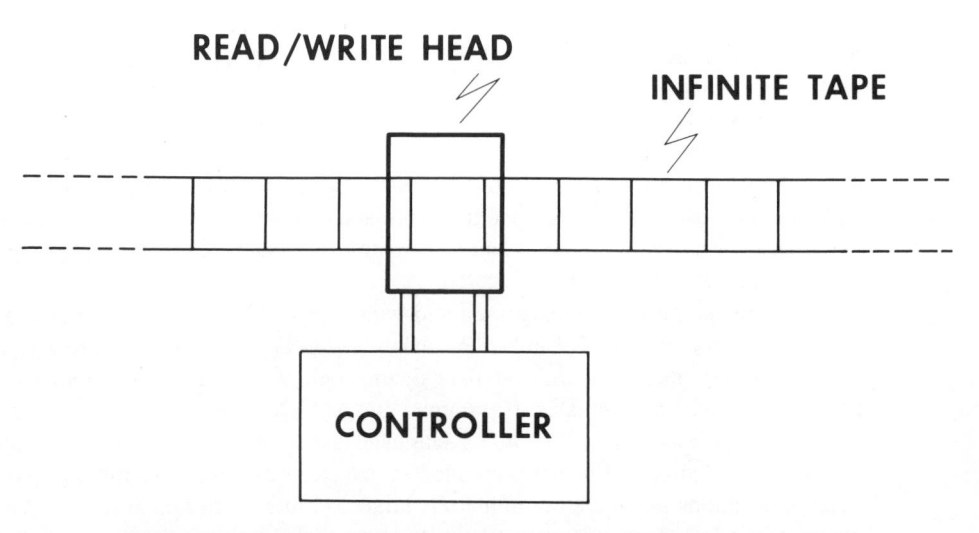

Figure 2-2 Elements of a Turing machine. (Source: Mark S. Monmonier, "Statistical Maps, Algorithms, and Generable Behavior," *Proceedings of the American Congress on Surveying and Mappng,* 36th Annual Meeting [1976], p. 409. Reprinted by permisson of the publisher.)

ALGORITHMS AND PROGRAMMING

Although the essence of computer-assisted cartography is in the algorithm, not the program, the steps outlined in an algorithm cannot be communicated to a computer without a more specific, rigidly phrased program. Communication between person and machine requires a programming language for the convenient and precise transfer of thoughts expressed in human vocabulary and mathematical symbols to instructions coded in bits as electrical charges or magnetic directions.

Programming Languages

Like English, French, and ancient Aramaic, programming languages have a grammar and vocabulary. This vocabulary rarely consists of more than 50 fundamental commands. Mathematical symbols such as $=$, $+$, $/$, and $>$ might represent common arithmetic and logical operations, and commands for operations not customarily represented by symbols are spelled out in the vernacular of the programmer, as for example, PRINT, GO TO, and REWIND. Thus there are English, French, German, and other versions of high-level languages such as FORTRAN and ALGOL. Names of arrays and unsubscripted variables can usually contain any available letters and numbers, although syntax usually dictates that the first character be a letter.

Programmers must observe a highly rigid grammar because computers are usually not designed to guess the intended meanings of vague or faulty commands and expressions. Minor errors in syntax that a programmer might easily overlook in a final inspection of a deck of punched cards will cause a batch-oriented computer to produce little more than a listing of the coded instructions and some (often cryptic) diagnostic messages. Interactive processors, which permit errors to be corrected and computing to be resumed, are preferred to batch processors, which are efficient for executing proven programs but cumbersome for developing new programs.

Computers use special programs to translate the programmer's code to binary machine code. An *interpreter* converts a program to machine code, instruction by instruction, as the program is executing, whereas a *compiler* translates all instructions as a group before the machine-code algorithm is executed.

Perhaps the most frequently used and widely available computer language is FORTRAN, an acronym for FORmula TRANslation. FORTRAN is oriented toward mathematical calculations for scientific, statistical, and engineering applications. Other scientific languages similar in syntax and grammar are ALGOL, for ALGOrithmic Language, developed in Europe, and BASIC, a language used widely in interactive computing. COBOL, for COmmon Business-Oriented Language, was developed for accounting and other applications manipulating numerous large, complex data files. PL/I, for Programming Language I, is an attempt to combine many of the desirable features of FORTRAN and COBOL. APL, for A Programming Language, is an interactive problem-solving language using algebraic and other symbolic notation for the compact coding of vector, matrix, logical, and other data-processing operations. PASCAL, introduced as an ideal "first language" for programmers, is an interactive language designed to aid the detection and correction of logical errors. It is a favorite of many computer-oriented industrial designers.

High-level languages provide the programmer with built-in functions for evaluating frequently used mathematical functions such as sines, cosines, logarithms, and square roots. A program statement with the expression SQRT(A) tells the compiler to generate the machine code needed to compute the square root of A, with as much accuracy as appropriate for a single-precision variable. The computer uses several steps to find the square root by successive approximation: a trial value is estimated for $A^{1/2}$ and squared; the result is compared with A, and a new trial value is estimated so that the next squared value will be closer to A. This process is repeated in a loop until the difference between A and the squared approximation of A is negligible.

The programmer can write his or her own functions and subprograms to be called at appropriate points in the main programs, or even by other subprograms and functions. Division of a complicated process, such as making a contour map, into convenient subprogram modules facilitates testing and *debugging* (finding and correcting errors), reduces redundant code, and provides program units that might easily be used efficiently in other main programs.

Manufacturers of plotters, CRTs, and other display hardware customarily provide or lease graphic software packages with subroutines that enable the programmer to draw complex pictures with relatively few program instructions. These subprograms, functions, and commands range from such primitive instructions as PLOT(X, Y), which tells a plotter to move to point (X, Y) with the pen down, to CIRCLE(X, Y, RAD, DEV), which generates a circle of radius RAD, centered at (X, Y), and approximated by short chords that lie no farther than distance DEV from the true circumference. Other commands might be called to change the scale of a map, plot lettering and special symbols, and draw axes for accompanying graphs.

Advanced interactive graphics systems often provide special functions such as *windowing* and *clipping,* which permit the selection of a rectangular portion, or window, of a larger picture and generate new intersection points for linear features that are clipped at the boundary. Command keys on a keyboard or functions in a menu grid on a digital tablet can facilitate the selection of these graphic operations. In contrast to programming for batch computing, interactive analysis and modeling is commonly done in *real time,* and the operator can easily alter his or her mental flow chart in response to the map appearing on the CRT.

In many respects it is reasonable to consider as a programming language the instructions that must be prepared by the user of a large "canned" mapping software package such as SYMAP or SURFACE II. To be sure, the user is actually providing data to a program written in FORTRAN or another high-level language; yet these data also have a vocabulary and grammar that are stated and enforced through the program's own diagnostic messages. Perhaps the only significant distinction is that a programming language in the usual sense has an associated compiler, assembler, or interpreter for generating the necessary machine code, whereas the software package contains all the machine code needed to comply with valid user instructions.

One other language must be mentioned, the control language of the executive system, monitor, or supervisory program that serves as an intermediary between the user and the computer. Large time-sharing computers require an operating system to regulate use of the computer and its peripherals, as do multiprogramming batch systems with several programs stored in different parts of internal memory for concurrent processing. The programmer must provide instructions in the control language indicating the compiler to be used, the tape, disk, or drum files for reading data and writing results, the display devices needed, and the account to which charges are to be billed. Control languages are often far richer in the variety of commands available and far more complex than high-level languages such as FORTRAN and COBOL. Consequently, systems programmers have emerged as a skilled priesthood to which applications programmers with enigmatic errors take their desperate pleas for help and enlightenment. Fortunately, the subset of control-language statements that are used frequently is comfortingly small, and the commands used with interactive computers are usually more akin to English than the cant of executive systems on batch processors.

An Example in FORTRAN

FORTRAN has been the most widely available and therefore the most transportable computer language. Most mapping software, especially the better documented packages, are written in FORTRAN. Although some hardware manufacturers have developed FORTRAN compilers with added features not available on other computers, a portable, rudimentary version of this high-level language, developed and supported by the American National Standards Institute, can be used worldwide on almost all computers with a FORTRAN compiler or interpreter. Because the grammar and vocabulary are relatively simple, FORTRAN command names can easily be translated from one nation's vernacular programming language to another by a straightforward translation program.

    The algorithm presented as a flow chart in Figure 2-1 provides a convenient introduction to FORTRAN coding. These instructions, which add up a series of N numbers stored in one-dimensional array X and store the result in variable SUM, can be coded as the subroutine ADD, to be called and executed by a main program.

```
        SUBROUTINE ADD (X, N, SUM, NDIM)
        DIMENSION X (NDIM)
        SUM = 0.0
        I = 0
   10   I = I + 1
        SUM = SUM + X (I)
        IF (I .LT. N) GO TO 10
        RETURN
        END
```

The first statement names the variables to be shared by the subroutine and the main program. X is a simple array with NDIM positions, N is the actual number of values of X to be totaled, and SUM is the result. The values of X, N, and NDIM are passed from the main program when the subroutine is called. A value for SUM is returned when the subroutine has been executed and the main program resumes control. The DIMENSION statement, used to declare the size of the array, receives NDIM, the maximum number of storage positions reserved for X, from the main program as a variable dimension; thus the size of the array can be altered without changing the code for the subprogram.

Statements 3 through 6 are executable commands that make the algorithm operational. The accumulator SUM and the counter I are cleared to zero. Then, in succession, I is incremented by 1 and the Ith value of X, expressed as X(I), is added to the current value of SUM. A repetition, or *iteration,* of these two steps is indicated by the next statement, which sends control to the statement labeled 10 if I is still less than its maximum value, N. Otherwise the computer will execute the following statement and return from the subroutine to the main program. The final statement merely marks the end of the subprogram for the compiler.

FORTRAN permits a more compact coding of program loops. Hence the subroutine can be written in seven statements, instead of nine, with a DO-loop replacing the initialization and incrementation of index I and the comparison of I with N.

```
        SUBROUTINE ADD (X, N, SUM, NDIM)

        DIMENSION X (NDIM)

        SUM = 0.0

        DO 20 I = 1, N

20      SUM = SUM + X (I)

        RETURN

        END
```

This compact subroutine is highly flexible; by altering N, and possibly NDIM, the subroutine can add any list of numbers.

A main program must declare a specific size for array X, set NDIM to a corresponding value, designate N, read the values of X from cards or some other data file, call the subprogram, and report the resulting sum. The following FORTRAN code reads both N and X from punch cards using FORMAT statements to indicate that (1) N is an integer to be found in the first five columns of the first data card, and (2) the values of X each have five decimal places and occupy the first ten columns of subsequent cards.

```
        DIMENSION X (20)

        NDIM = 20

        READ 1, N
```

```
1          FORMAT (15)

           DO 10 I = 1, N

10         READ 2, X (I)

2          FORMAT (F10.5)

           CALL ADD (X, N, SUM, NDIM)

           PRINT 3, N, SUM

3          FORMAT (' THE SUM OF ',I5,' NUMBERS IS ',F12.5)

           STOP

           END
```

Instructions to read and write data can be complex, but the beginning programmer might find comfort in FORTRAN's use only as a written, not a spoken, language.

FORTRAN programs are seldom used to add only 10 or 20 numbers, and surely the previous exercise would be an inefficient use of the capabilities of a large computer. Many programmers who code large programs in FORTRAN find a second language such as APL convenient for simpler tasks. At a remote terminal, APL might use only three statements to enter and add a few numbers and print the results.

$$X \leftarrow 7\ 8\ 12\ 2\ 4\ 5\ 15$$

$$SUM \leftarrow +/X$$

$$\text{'THE RESULT IS ', SUM}$$

For a less elegant result, an even more compact form is possible.

$$+/7\ 8\ 12\ 2\ 4\ 5\ 15$$

If many computations are required, as is common in mapping programs, waiting for results from interactive computing can be tedious when many other users need to be served as well. Moreover, the amount of internal memory usually available to any single interactive user is commonly restricted. Time-sharing systems with *virtual memory*, short for "virtually unlimited memory," are less likely to have as severe memory constraints; program and data are divided into a number of *pages*, only some of which are kept in internal memory at any time. The process is said to be "transparent to" the user, who is unaware of the shifting of pages between internal and external memories.

Mapping on the Line Printer

A simple algorithm illustrates how even a computer with only a line printer or remote terminal for data display can be programmed to make maps. Although the maps produced by the procedure described here are neither elegant nor esthetically pleasing, line-printer

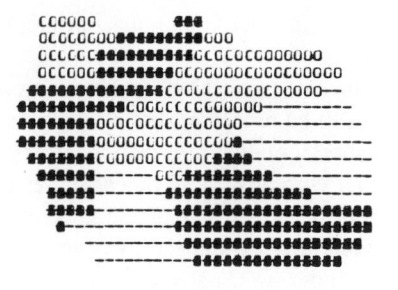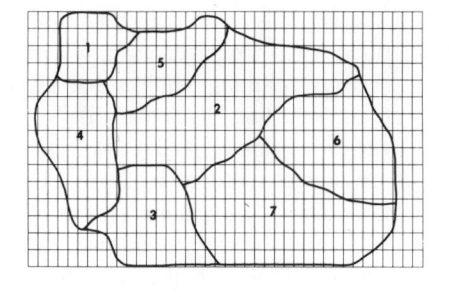

Figure 2-3 CMAP choropleth map (left) and line-printer grid superimposed on areal unit boundaries (right). Areas 3 and 6 have been assigned the lightest shading, areas 1 and 2 an intermediate shading, and areas 4, 5 and 7 the darkest shading. (Source: Mark S. Monmonier, "Internally-stored Scan-lines to Increase the Efficiency of Small-memory Line-printer Mapping Programs," *The American Cartographer*, 2, no. 2 [October 1975], 145–53. Reprinted by permission of the American Congress on Surveying and Mapping.)

maps can display geographic distributions to provoke ideas, test hunches, or preview map patterns before a more expensive and visually appealing map is drafted.

CMAP was developed by Morton Scripter for the U.S. Bureau of the Census to produce at a predefined scale inexpensive maps of statistical data for areas on computers with small memories.[1] CMAP produces maps called *choropleth* maps, from the Greek words *chóros,* for open area, and *pléthos,* for fullness or magnitude; a single area-shading symbol covers each areal unit, which thus appears to have the same mapped value throughout. The printer grid of rows and columns of printable characters must be apportioned among the various census blocks, tracts, counties, states, or countries of the region to be mapped (Fig. 2-3). The range of data values for the map must be divided into intervals or classes so that the computer can assign each areal unit one of a small number of area symbols. These shading symbols are graded according to the tonal intensity of their alphanumeric characters, with a very light character such as the minus sign representing the category of lowest values, and a relatively dark character, perhaps the pound sign (#), portraying the interval with the highest values. Map viewers can readily distinguish among six or seven area shadings at most. Choropleth maps are thus generalized in two ways: (1) in the geographic resolution inherent in the size of the areal units employed, and (2) in the aggregation of different but similar data values into mapping categories.

Alphanumeric characters produced by standard line printers do not fill the entire grid cell; an uninked halo even surrounds the customarily rectangular letter O. Although this mesh of uninked dividing lines is mandated, most printers can produce darker cell interiors by overprinting, say, the letters O, X, and I, and the minus sign. The *gray scale* of these

---

[1]Morton W. Scripter, "Choropleth Maps on Small Digital Computers," *Proceedings,* Association of American Geographers, 1 (1969), 133–36.

area symbols can be extended farther toward solid black if an impact printer uses a fresh ribbon or if the paper yields a carbon copy.

Overprinting requires that successive lines be printed before advancing the carriage to the next line. Thus each row of the map can be processed and printed independently, and the computer need retain in internal memory only the information needed to describe the geometry of the single row, or *scan line*, currently being processed. Each scan line is divided into *segments*, strings of contiguous cells representing the same data area (Fig. 2-4). The various geographic zones are identified by integers between 1 and M, the number of such areal units. This numerical indexing usually follows the alphabetic sequence of area names. Blank, background cells representing territory outside the mapped region are also numbered, as area $M + 1$. Each scan-line segment can be identified by the position inward from the left of its rightmost column and the index number of the areal unit, and the entire scan line can be coded by adding the number of segments to these successive pairs of rightmost columns and index numbers (Fig. 2-4).

The general structure of the CMAP algorithm is quite simple, as shown in Figure 2-5. The corresponding program needs the following data:

1. A parameter card containing the numbers of areal units, categories, and scan lines, as well as the width in columns of the printer grid.

2. A card containing a title to be printed above the legend.

3. A format card describing, according to FORTRAN rules, the arrangement on individual cards of the data values for the areal units.

4. A card containing the class breaks used to divide the data range into mapping categories.

5. For each category, a card containing the overprint symbols for area shading.

6. A deck of cards containing the data values.

7. A deck of cards defining the scan lines.

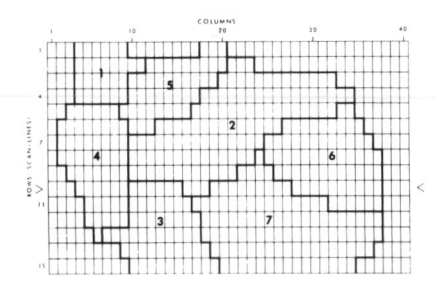

Data for seven scan-line segments for row 10

Segment	Area Identification Number	Rightmost Column
1	8 [background]	3
2	4	9
3	3	15
4	2	18
5	7	27
6	6	37
7	8 [background]	40

Figure 2-4 Printer grid (left) divided into scan-line segments, with coding shown for row 10 (right). (Source: Mark S. Monmonier, "Internally-stored Scan-lines to Increase the Efficiency of Small-memory Line-printer Mapping Programs," *The American Cartographer*, 2, no. 2 [October 1975], 145–53. Reprinted by permission of the American Congress on Surveying and Mapping.)

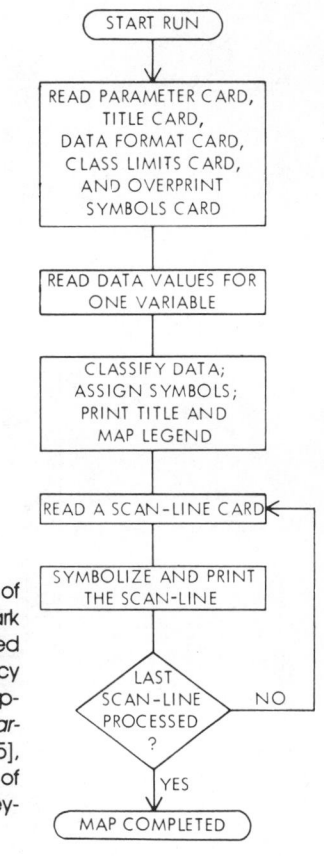

Figure 2-5 Simplified flow chart of the CMAP algorithm. (Source: Mark S. Monmonier, "Internally-stored Scan-lines to Increase the Efficiency of Small-memory Line-printer Mapping Programs," *The American Cartographer*, 2, no. 2 [October 1975], 145–53. Reprinted by permission of the American Congress on Surveying and Mapping.)

The data values are then read into internal memory. Each value is compared against the class breakpoints assigned to a category, and assigned the overprint symbols for that category. After printing the title and legend, the computer processes one scan line at a time, until all rows of the map have been printed.

Two procedures merit a more detailed examination, classification of the data and symbolization of the scan lines. The classification algorithm is the simpler process (Fig. 2-6). An outer loop with index I assures that all M data values are classified, and a single interior loop with index J compares each data value X(I) with the upper limit ULIMIT(J) of the P categories for which the user has provided intervals and symbols. Before any comparisons with upper limits are permitted, however, each data value is compared with the lower limit XLOW of the first class. If X(I) is less than XLOW, or if X(I) is not less than any of the P upper limits, the value is outside the range of the data, and the four overprint symbols in the row I of two-dimensional array ISYMB receive the respective four characters stored in the four rows of one-dimensional array IOUT. Otherwise, X(I) is found to be less than one of the progressively higher upper limits ULIMIT(J), indicating that area I belongs to category J. The appropriate four overprint characters are then

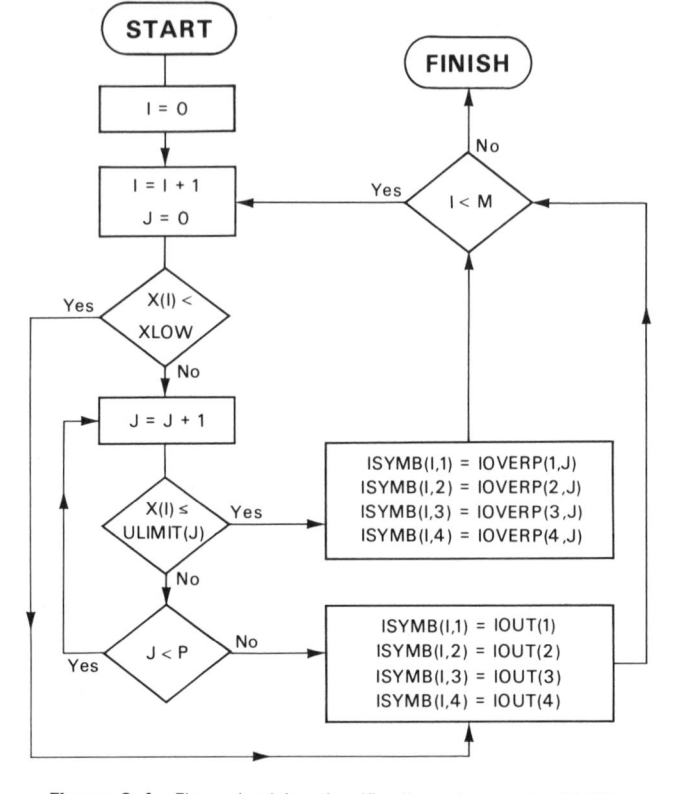

Figure 2-6 Flow chart for classification of areas for CMAP.

transferred from row J of two-dimensional array IOVERP to row I of array ISYMB. Although represented in memory by strings of binary digits, the alphanumeric characters are converted for printing to the corresponding letter, number, or special symbol.

If these symbols and flow chart seem confusing, a large sheet of paper can be used to simulate the program's directions to the computer. To understand this process more fully, begin by drawing simple boxes to represent the storage locations of variables I, J, M, P, and XLOW. Also draw stacks of boxes to represent one-dimensional arrays X, ULIMIT, and IOUT, and grids to represent two-dimensional arrays ISYMB and IOVERP. Then develop sample data for M, P, X, ULIMIT, IOUT, and IOVERP, and simulate the computer by following the flow chart (Fig. 2-6) step by step and around and around until every circuit through the outer loop with index I is completed. Although this exercise might increase the time in which you planned to read this chapter, the added understanding of computer processes should justify the effort.

The symbolization algorithm shown in Figure 2-7 combines the data describing the scan-line segments with the printing characters stored in array ISYMB for the M areal units. This procedure uses four loops, nested one inside another. The outer loop, with index K, assures that the array LINE is loaded with the appropriate overprinting characters for each of the N rows of the map. The second loop, with index L, processes each of the NUMSEG segments of a scan line. For each segment the place index number INDEXP(L) and the number of the rightmost column IRITEP(L) are read from external memory or a data-entry device. The third loop, with index ICOL, assures that each column of array LINE receives printing characters. Unlike the counters for the other loops, ICOL is not reset to 0 at the outset of every series of iterations: for each segment, ICOL refers only to additional columns and thus moves only to the right in array LINE until the rightmost column IRITEP(L) of the segment has been treated. The most interior loop, with index J, assures that printing characters are placed in each of the four rows of column ICOL of array LINE. If the place index INDEXP(L) equals N + 1, indicating a background cell in the map grid, all printing characters are blanks. Otherwise, the appropriate printing characters are obtained from list ISYMB containing overprint symbols for each of the M mapping areas. For the first line of the example map in Figure 2-3,

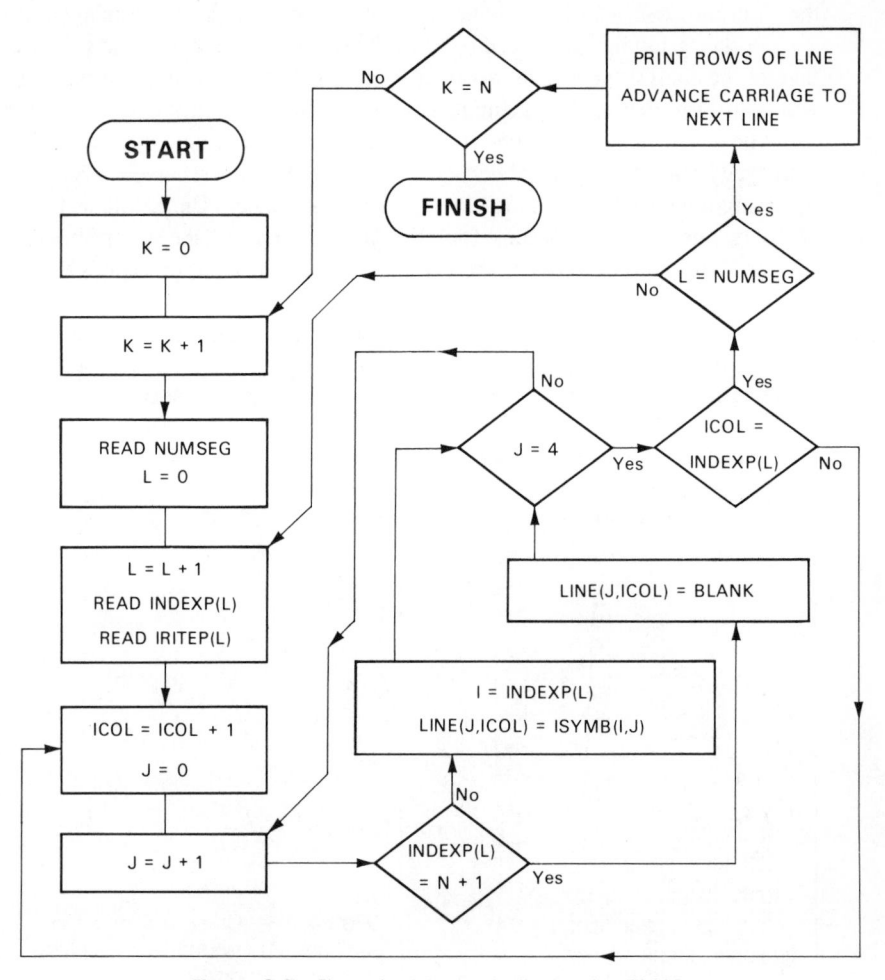

Figure 2-7 Flow chart for symbolization by CMAP.

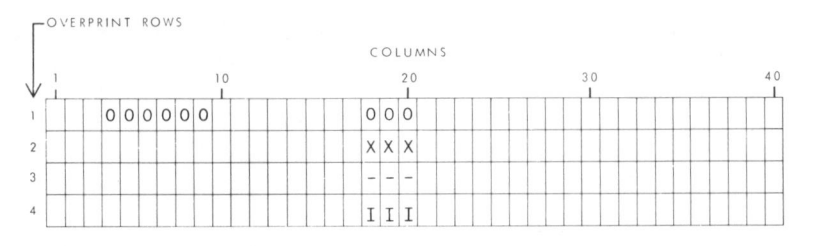

Figure 2-8 Overprint rows in array LINE for first printed scan line of map in Figure 2-3. (Source: Mark S. Monmonier, "Internally-stored. Scan-lines to Increase the Efficiency of Small-memory Line-printer Mapping Programs," *The American Cartographer*, 2, no. 2 [October 1975], 145–53. Reprinted by permission of the American Congress on Surveying and Mapping.)

the algorithm will load LINE with the four rows of overprint lines shown in Figure 2-8.

Printer-grid maps produced by CMAP are less impressive to the skilled cartographer than to the analyst who might appreciate more readily the need to display, inexpensively and with little delay, the essential geographic characteristics of a large number of distributions. The researcher interested in a region with numerous areal units might be dismayed because useful local details are revealed only if a relatively large map is prepared by taping together two or more strips of computer paper. Unlike other, necessarily more complex line-printer programs for choropleth mapping, the comparatively unadorned CMAP algorithm does not permit a change in scale: if a larger scale is needed, the scan-line data must be recoded from a map with a larger scale. Photographic reduction of a relatively large CMAP to obtain a copy suitable for publication is possible, although lighting and exposure will be complicated. Linear features to provide a geographic frame of reference can be added on a transparent overlay, as can neat, crisp type for the map

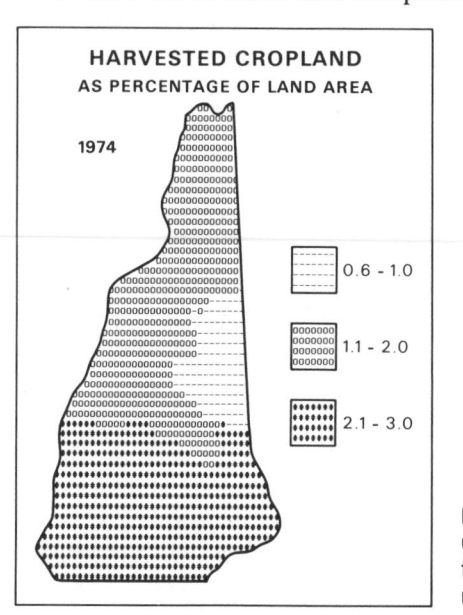

Figure 2-9 Crisp type and manually inked boundaries can improve the appearance of a line-printer map.

title and other labels (Fig. 2-9). If many choropleth maps are to be produced at the same scale and for the same set of areal units, implementation of a simple mapping system such as CMAP can be practical and cost effective.

Advanced Developments

Elementary principles such as those employed in CMAP apply as well to some advanced systems. Overprinting, for example, is used in more sophisticated line-printer software. Both the scan-line concept and a modification of overprinting are used in raster-mode color CRT displays. Three values for each scan-line segment represent the individual intensities of the three additive primary colors, blue, green, and red.

Other principles demonstrated by CMAP, such as the use of stored addresses coded for array subscripts, might change radically with the widespread adoption of improvements in computer architecture. A particularly promising development is the *associative storage* characteristic of a *content-addressable* memory. Locations in memory need not be computed from array subscripts by *fetch functions*. Instead, data items may be stored with associated identifying attributes, or *keys*, such as place names or land-cover types. Entries thus may be retrieved in groups by key rather than singly by address. Retrieval is a response to the query, "What in memory?", not "Where in memory?"

It is simpler, of course, if data entries can be processed by their attributes alone, without specific programming reference to individual storage locations. An advanced design especially useful for areally extensive data is the *array processor*, a machine that can work simultaneously, or "in parallel," on many storage locations. As discussed in the next chapter, geographic data can be expressed as superimposed layers registered to the same grid system. Each cell represents a rectangular part of the Earth, and each layer consists of the values for a particular land attribute. Array processors permit the rapid storage, retrieval, and analysis of gridded raster data.

Hardware improvements will yield many software improvements. From the user's point of view, computer languages will become simpler. FORTRAN, as illustrated in the CMAP example, is a useful and generally straightforward language for a variety of scientific applications, including mapping. Yet with the accumulation of ever-greater masses of digital cartographic data, a still more useful computer language will be provided by the data-base management system (DBMS). Data-base languages will simplify file management and retrievals far too cumbersome for standard programming languages.

Data transmission is changing as well. As presented here, CMAP reads punch cards, the standard storage medium during the 1960s and early 1970s for small amounts of data. Card decks were commonly carried by hand to the computer center before the adoption of time sharing and office and home terminals. Magnetic tapes were used for larger files, particularly when data were sent through the mail or by air express. Data now are commonly transmitted through telecommunications lines: over telephone wires, by microwave relay, and by satellite. With optical-fiber transmission, data can be exchanged even more rapidly.

A natural consequence of highly efficient data transmission is the *distributed data base*. Geographically dispersed computers linked by a data network can eliminate redundancy and permit various components of a data base to be managed locally by subject-matter experts fully aware of what the digital data represent. Data can be dispensed only to those who have the need, skill, and permission to use them.

With the extensive use of data networks, terminals, and home computers, more employees might well be working at home or in hotel rooms, if traveling. Inexpensive, uncomplicated, portable terminals that promote computer conferencing and time-share programming and data analysis most likely will not provide high-resolution color graphics. For these users, the simple scan-line CMAP algorithm will be quite helpful.

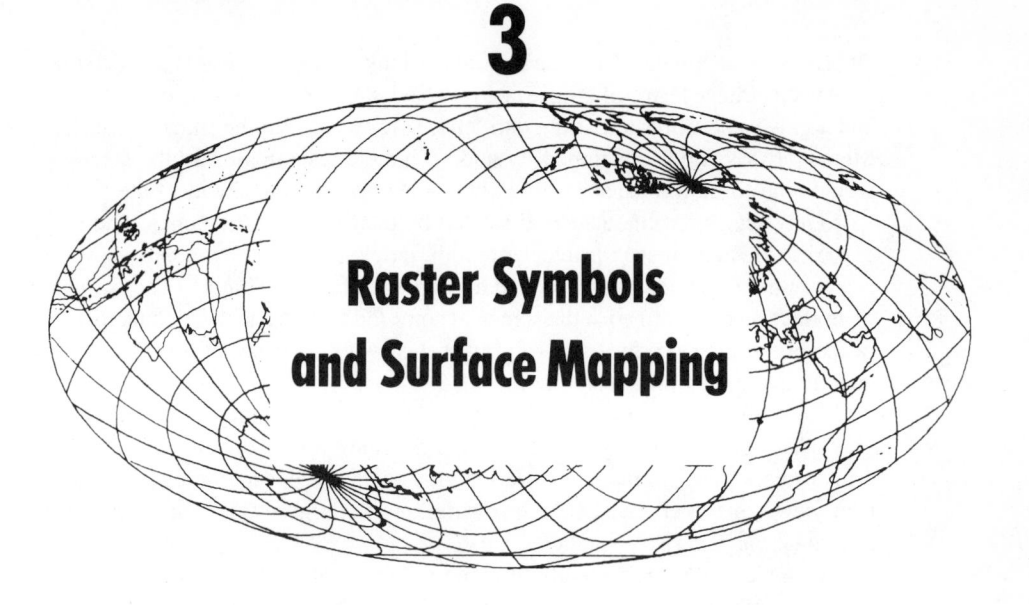

3

Raster Symbols
and Surface Mapping

Raster symbols should not be confused with raster data. "Data" is derived from the Latin verb *dare*, meaning "to give." Data frequently must be accepted as given, whereas symbols can be more freely chosen. Although raster data are more directly amenable than vector data to portrayal by raster display devices, such as the line printer and refresh CRT unit, vector-to-raster and raster-to-vector conversions before display are common. This chapter demonstrates the use of vector, point-coordinate data with a surface interpolation algorithm to produce raster-symbol contour and choropleth maps. The SYMAP program, a mapping software innovation in the late 1960s but long since surpassed in efficiency and esthetics, is used to illustrate several important principles. Although plotter software produces maps more visually pleasing than the coarse, line-printer SYMAPs, improved raster processors and display units promise a renewed importance for raster symbols. The chapter begins with a discussion of the cartographic uses and drawbacks of raster symbols.

USES AND LIMITATIONS OF RASTER SYMBOLS

Of the variety of characteristics of raster symbols, perhaps the most important is tone density, the position on the gray scale from pure white to solid black of the graphic density of the basic raster unit, the *grid cell*. A gray scale can be *complete*, and include a solid symbol filling the grid cell, or *truncated*, with the darkest symbol less dense than solid black. Gray scales might be truncated because the image area, for example, the alphanumeric symbols printed on the standard line printer, does not extend to the perimeter of the cell. Images can, of course, be reversed photographically as well as electronically, and a gray scale truncated at the dark end, if reversed, will be truncated at the light end. Moreover, a symbol developed by adding light, for example, on the screen of a CRT, would tend to be truncated at the light, rather than the dark, end of the gray scale.

Gray scales are characterized as well by their range of graytones. The very fine grid of the electrostatic matrix printer or the ink-jet plotter has a *dichotomous* gray scale: each cell is either darkened or blank, and all darkened cells look alike. The variable intensity of an electron beam aimed at the screen of some CRT units and the alphanumeric characters of the line printer produce *varied* gray scales, with a variety of tones that can represent quantitative geographic differences.

Resolution and map scale are important in differentiating dichotomous from varied gray scales. Differences in pattern spacing among area symbols produced on a matrix printer might yield a graphically varied gray scale if reduced photographically. What the map viewer may perceive as pattern variation at a larger scale will usually be perceived as a tonal variation at a smaller scale.

Graphic marks in the cells of some raster-mode displays such as color CRT units can be colors rather than shades of gray. The user controls not only the value, or brightness, of a color, but also the hue, or wavelength, and the chroma, or degree of saturation. The graphics system might provide a standard color *pallet* with, perhaps, 64 different colors chosen by the system designer. Each color would be represented by a unique six-bit code. The user achieves appropriate contrasts and gradations by assigning color codes to the various features portrayed on the screen. Colors are related to features in a *look-up table*.

The individual marks used to lighten or darken the cell may be geometrically simple, such as a circle, square, or hexagon, or graphically complex, as are the letters and numbers of the standard line printer. Although never widely used, a print train with a variety of simple geometric forms, such as circles and hexagons, of various sizes can produce, without overprinting, esthetically pleasing area symbols with a variety of tonal and pattern differences (Fig. 3-1).

Equal in importance to the gray scale is the spatial resolution of raster symbolization. Resolution determines the sensitivity of a symbol for graphic detail and usually is measured as the length of the side of a square cell. Cells may be elongated, of course, as is common with line printers, which are designed primarily for tabular displays. Cells may also have a variable width, in the sense of scan-line segments for the CMAP program and a raster-mode CRT unit, yet a minimum practical cell width, the width of the narrowest possible alternating light and dark segments, for example, can be recognized. The resolution of raster displays is commonly stated in scan lines per unit length.

The resolution of a CRT display is usually determined by the amount of memory devoted to the display and the size and quality of the tube, so that resolution is generally proportional to cost. At the low end of the scale are coarse CRTs with 40 by 40 grids offering noticeably blocky graphics. Moderately priced CRTs designed primarily for graphics have grids with between 256 by 256 and 512 by 512 cells. Higher resolution is more expensive, with a screen of 1,024 by 1,024 cells requiring four times the memory needed for a display with 512 by 512 cells. The central processor must also be significantly

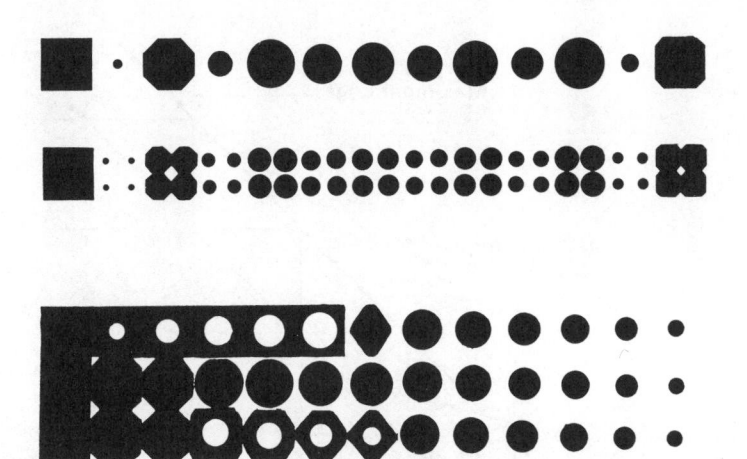

Figure 3-1 Enlargement of special symbols developed for line-printer chain to provide fuller range of consistent gray tones for area shading. (Courtesy Kurt Brassel.)

faster if display and processing times are not to be markedly slower. Large-screen CRT systems with high resolution thus await further improvements in memory and CPU design.

Raster displays can be compared in resolution by computing the change in scale needed to obtain a comparable degree of detail. The grid of a matrix printer plotting 70 lines/cm can be replicated on a line printer with square grid cells 0.25 cm (0.1 in.) on a side, but the map would have to be reduced photographically, in stages, to 5.7 percent of original scale after the strips of printout were assembled into a large mosaic. The assembly and photographic reduction of this mosaic are awkward at best, because the display scale is much greater, 17.5 times in this example, than the scale of a map with comparable spatial resolution produced by the matrix printer.

This concept of excessive enlargement not only is convenient for describing the graphic capability of display hardware, but it also helps to explain why raster displays are not normally preferred for symbols with intricate linear elements. Suppose, for example, an experimental study of the "toothiness" of the edges of a diagonal line 0.02 cm (0.008 in.) thick determines that map viewers generally do not object to differences between minimum and maximum line widths less than 0.01 cm (0.004 in.). The maximum difference permitted on each side of the line is thus 0.005 cm (0.002 in.), a degree of serration that requires cells no larger than 0.007 cm (0.003 in.) on a side (Fig. 3-2). This resolution would be attained by matrix printers plotting at least 140 lines/cm (360 lines/in.). A map prepared on a line printer with an original resolution of 4 lines/cm (10 lines/in.) would require photographic reduction to only 2.8 percent of the initial scale to provide a map comparable in scale and resolution. In this example the excessive enlargement factor of 36 clearly would rule out photographic reduction from a line-printer mosaic. Experiments with human subjects might suggest a more tolerant attitude toward the obviously serrated lines of most raster displays, but even a maximum tolerable difference in line width of 0.05 cm (0.02 in.), five times greater than the hypothetical value used

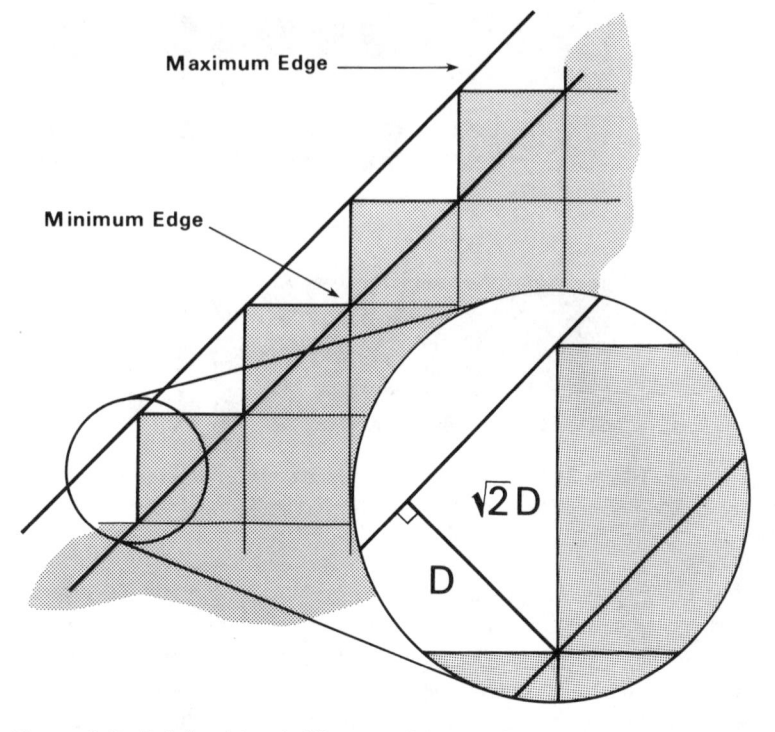

Figure 3-2 Relationship of difference D between maximum and minimum edges of a serrated diagonal line and minimum raster cell side $\sqrt{2}\,D$.

here, would require a display with at least 28 lines/cm (72 lines/in.), thus excluding most CRT units and all standard, impact line printers.

The preceding illustration also explains why the coarser raster display devices are seldom ideal for graduated point symbols such as circles, pie charts, and pictorial symbols scaled so that the apparent areas of the symbols represent magnitudes for cities, industrial plants, and other point locations. Simple square or rectangular point symbols, with only vertical or horizontal sides, are more practicable for raster display than circles and other forms with either intricate or smooth, curving edges. Raster displays with poor resolution, such as the line printer, are limited further in both the precision with which point symbols can be positioned and the accuracy with which subtle differences in magnitude might be portrayed. If the point phenomena are clustered, as is common for urban centers and economic activities, raster point symbols are likely to be visually awkward unless the spatial resolution is quite fine. The dark intersecting circular arcs of clustered graduated circles, with empty or lightly shaded interiors, are plotted more easily by most vector display devices and can communicate effectively despite moderate amounts of overlap.

Resolution might also limit the efficiency of raster displays for making dot-distribution maps. Cells covered completely or largely with solid black might serve as dot symbols, each representing some stated number of people, dairy cattle, or dilapidated houses, but the raster grid constrains not only the accuracy with which dots can be positioned but also the range of dot densities. Dot-distribution maps are intended to communicate differences in density, an objective that might be inhibited if close dots must be either side by side or one or more cells apart. A well-designed dot map depicts the greatest mapped density with dots that barely coalesce, a visual effect requiring a fine raster grid so that each dot occupies not a single cell but a circular group of many cells. Needless to say, vector display of dot distributions is inherently less complex.

Raster symbolization is more appropriate for portraying quantitative differences among areal units than for most other cartographic tasks. Tonal variations in area symbols are promoted by display devices with gray scales that are both complete and varied. Display units with dichotomous grid cells can, of course, simulate the visual effects of a wider range of graytones if square-shaped groups of contiguous cells on the display are assigned to a single cell of the data. Lighter tones thus can be obtained by reducing the number of these blackened subcells, and if the display grid is fine grained, the visual effect of tonal, or value, differences will dominate a possible impression of pattern or textural differences. This distinction between textural and tonal variations is particularly significant for line-printer maps, the appearance of which usually is improved by a small amount of photographic reduction, say, to 50 percent of the original linear scale. Alphanumeric printer symbols, particularly with overprinting, can effect tonal contrasts suitable for viewing from a distance, but photographic reduction generally is needed to suppress the visual dominance of the row-and-column texture of the space between characters, as well as the visual intrusion of the separate identities of individual letters and numbers.

Volumetric distributions such as population density and terrain surfaces frequently are represented by isoline, or contour, maps. Substantial photographic reduction from a greatly enlarged original line-printer map would be needed to maintain the fine details characteristic of topographic contours. Nonetheless, a raster display unit with a relatively coarse grid can portray effectively a contoured surface if area-shading symbols emphasize the regions between contour lines, rather than the often erratic form of thin contours. The map reader thus can estimate elevation categories from the graytones of quantitative area symbols instead of reading specific elevations directly from labels on the contours. Excessive enlargement and subsequent photographic reduction might, of course, be required for the detailed portrayal of irregular surfaces. Moreover, a legend must be provided for the area symbols, and vertical detail is limited because normally no more than eight visually distinct graytones are possible. The accuracy of a vector-mode isoline map might be greater because many more contour lines can be included, but the strong visual expression of a generalized isoline map symbolized with the coarse raster-mode graytones of the line printer might be more suited for many cartographic needs.

Among the most visually effective qualitative symbols are area symbols differing in hue. Raster display units, such as color CRTs and color electrostatic dot matrix printers, are particularly useful for mapping qualitative phenomena such as land-use categories. Color can also be highly effective on two-class quantitative maps, for example, a map differentiating the more affluent parts of a city from census tracts with below-average family incomes. Color CRT units are particularly useful for interactive displays and might well replace the present monochrome CRTs to the same extent that color television, rather than black-and-white television, has become the norm for home viewing. Color separations produced by COM units and photohead flatbed plotters can be used to make printing plates for offset lithography and thus are suited to the computer-assisted preparation of road maps and a variety of atlases.

The processes of image transfer (photographic exposure and development, the etching of printing plates, and the production of multiple copies by lithographic or letterpress printing) rarely preserve exactly the thickness of image elements. Thin or small image elements such as the dots of an area-tinting screen are particularly vulnerable to severe image growth or shrinkage when an enlarged map must be photographically reduced to achieve finer resolution for the final map. The practicable range of photographic scale change is limited in particular by the method of reproduction, and the designer of computer-produced maps must consider the full series of processes intervening between the plotted map and the graphic seen by the map user. When color reproduction is involved, a combination of vector and raster displays might provide a more realistic solution than either mode alone. During reproduction, vector symbols for point and linear features can be merged with raster symbols portraying quantitative area differences.

RASTER-MODE SURFACE MAPPING

Among the earliest computer-produced maps were raster-mode representations of statistical and terrain surfaces displayed on the line printer. These maps, although crude in appearance, are still adequate for many purposes, including wall displays to be annotated with felt-tip marking pens. SYMAP, a flexible mapping program designed for the line printer, illustrates a somewhat primitive but widespread form of computer-assisted cartography. The program might be called obsolete, but many of the raster-mode principles are valid nonetheless.

The SYMAP Program

SYMAP was the brainchild of Howard T. Fisher, who in the mid-1960s joined the faculty of Harvard University, where he became Director of the Laboratory for Computer Graphics and Spatial Analysis. Fisher's background was in architecture and planning, and the terminology he used to describe the features of his program were coined to reflect a functional view of mapping, rather than traditional cartographic principles. Choropleth maps, for example, were called "conformant maps" and mapping classes were called "levels." The name SYMAP is an acronym for "SYnagraphic MAPping," a term intended to mean "acting together graphically."[1]

Numerous modifications were made by the staff of the Laboratory, at that time an eclectic group of engineers, planners, designers, and computer scientists, but not car-

---

[1]James W. Cerny, "Use of the SYMAP Computer Mapping Program," *Journal of Geography,* 71, no. 3 (March 1972), 167.

tographers. The program was disseminated widely among university departments of geography, planning, and landscape architecture, thanks in no small measure to a substantial grant from the Ford Foundation, which sponsored several user conferences at Harvard. While professional cartographers, many of whom had the skills to develop a comparable program, pursued their academic research and governmental responsibilities, Fisher and his colleagues filled a void, competently although somewhat idiosyncratically.

SYMAP produces two principal types of map: choropleth and isoline. Whereas area polygons encoded as vector data provide the basis for choropleth maps, point data, including control points located at the centers of areal units, may be used to generate isoline maps. The principal type of isoline map represents three-dimensional surfaces with contours interpolated from elevations at nearby data points. These contours usually are represented as chains of blank cells running between shaded areas, each representing a range of surface values (Fig. 3-3, left).

If interpolation is based upon only the single data point closest to each grid cell, the resulting "proximal" map displays chains of blank cells outlining *Thiessen polygons,* which, by definition, enclose all cells lying closer to a data point than to any other data point. Each Thiessen polygon is shaded with the graytone of the mapping category containing the data value for the central point (Fig. 3-3, right). Unlike the contour map,

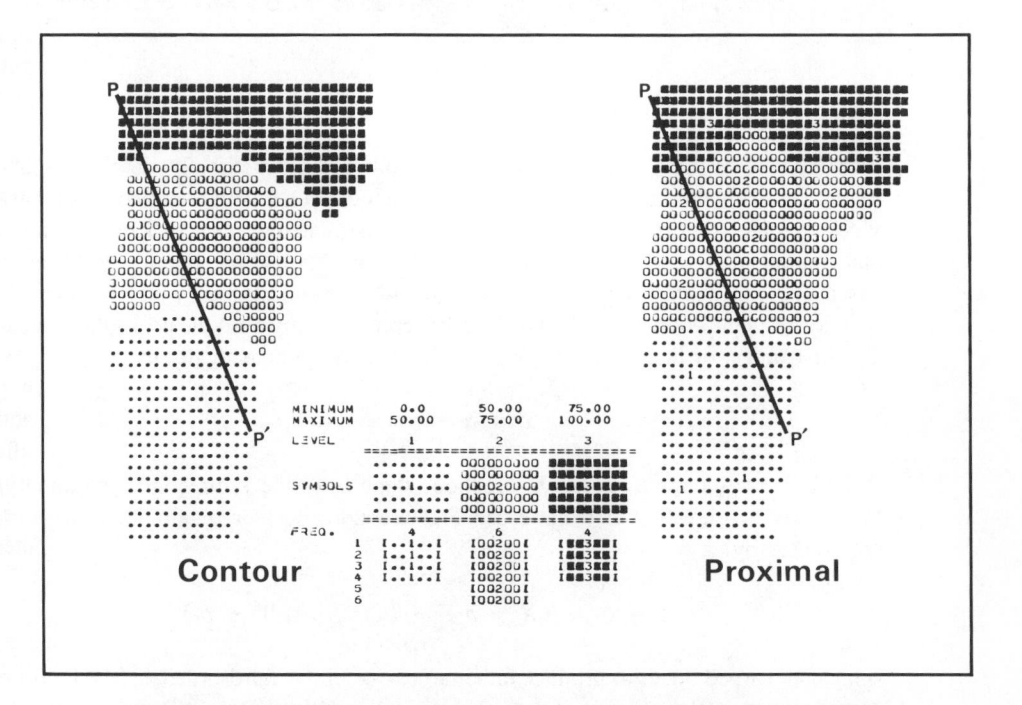

Figure 3-3 Contour and proximal maps produced with SYMAP for Vermont county center points, showing Canadians as a percentage of all foreign stock. Profiles for transect *P-P'* are shown in Figure 3-4.

a proximal map does not represent a continuous surface: the outlines of the Thiessen polygons represent vertical cliffs between plateaus or shelves at various categorized elevations. A vertical profile through a proximal map is stepped and discontinuous, whereas a similar profile through a contoured surface is smooth and continuous (Fig. 3-4). In this sense the proximal map is similar to a choropleth map.

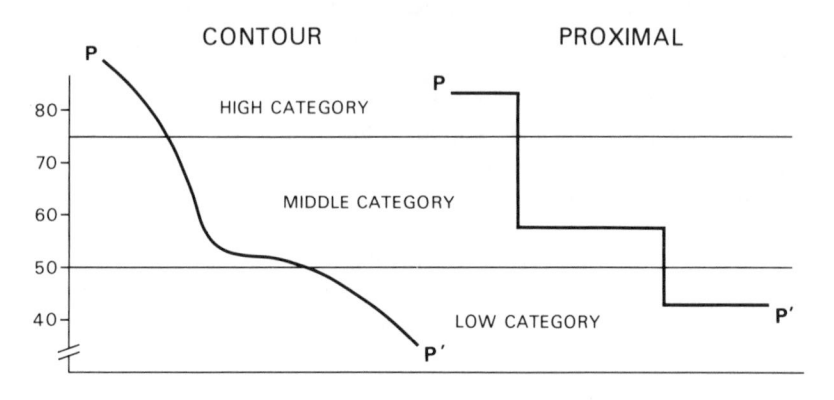

Figure 3-4 Profiles along transect *P-P'* for the contoured and proximal surfaces shown in Figure 3-3. Horizontal lines represent class breaks at 50 and 75.

Trend Surface Maps

SYMAP also displays another, generally smooth and geometrically simple type of contoured surface called a *trend surface*. This method of estimating surface elevations provides a *global* solution, based upon all data points treated at once, whereas the proximal and contour maps, discussed in more detail later in this section, provide *local* solutions, based only on nearby data points. A "low-order" trend surface with elevation Z as a function of plane coordinates X and Y can be represented by a simple polynomial equation. The first-order, or linear, trend surface, represented mathematically by

$$Z = a + b_1X + b_2Y$$

is a plane, with slope b_1 in the X direction, slope b_2 in the Y direction, and intersecting the Z axis at elevation a (Fig. 3-5). The second-order, or quadratic, trend surface, represented by

$$Z = a + b_1X + b_2Y + b_3X^2 + b_4XY + b_5Y^2$$

is a plane warped once, to produce either a peak or pit. A third-order surface is still more complex because the addition of the cubic terms X^3, X^2Y, XY^2, and Y^3 permits two extreme points, a peak and a pit. SYMAP also maps fourth-, fifth-, and sixth-order trend surfaces, with additional extrema provided by more terms in the polynomial equation.

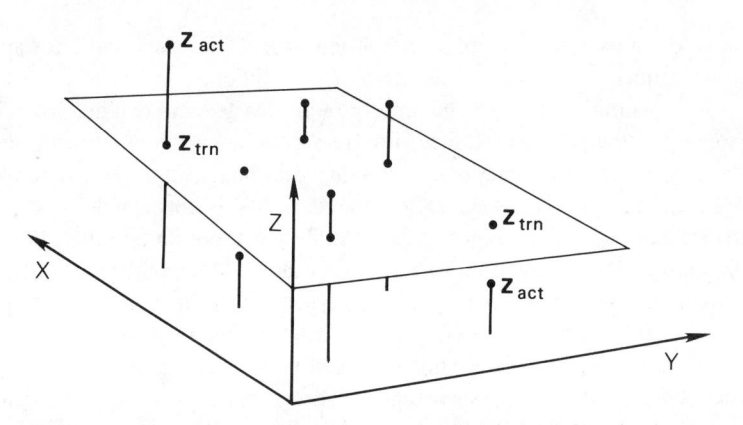

Figure 3-5 Perspective view of linear trend surface fit to six data points, four lying above the tilted plane and two below. Each point, whether above or below the trend surface, has an actual height Z_{act} above the X-Y plane and a trend height Z_{trn}, the elevation at the point of the planar surface.

A trend surface is fit to the data points by least-squares regression.[2] A low-order trend surface is constrained to a simple geometric shape and usually cannot pass through all the three-dimensional data points (Fig. 3-5). The surface thus represents an average regional trend in the data and commonly passes either above or below every data point. The vertical deviation at data point i is the difference between actual elevation $Z_{act}(i)$ and the elevation $Z_{trn}(i)$ computed for point $(X(i), Y(i))$ from the trend equation. The least-squares criterion states that squaring these vertical deviations for all data points and then adding up the squared deviations will yield a result as low as possible.

In mathematical terms, the least-squares criterion is expressed as

$$\sum_{i=1}^{n} (Z_{act}(i) - Z_{trn}(i))^2 = \text{minimum}$$

where $\sum_{i=1}^{n}$ means "sum the following expression for $i = 1$, for $i = 2$, and for all intermediate terms in steps of 1 through $i = n$." The terms $Z_{act}(i)$ are part of the data, and the terms $Z_{trn}(i)$ can be computed from the coordinates of the data points for a specified order of trend surface. A set of equations derived from calculus permits the calculation of the Z intercept a and the slopes $b_1, b_2, \ldots$ for the designated trend-surface polynomial (linear, quadratic, and so on) that satisfies the least-squares criterion. No other set of intercept and slope coefficients will yield a smaller sum of squared deviations than this "best-fit" surface. Because the equations for finding the least-squares solution are easily

[2]John C. Davis, *Statistics and Data Analysis in Geology* (New York: John Wiley & Sons, Inc., 1973), pp. 322–52.

derived mathematically and do not require successive trial-and-error approximations, the trend surface is said to have an *analytical* solution.

A quadratic surface is inherently more flexible and can better "fit" the three-dimensional data than a linear surface. Progressively higher order surfaces provide increasingly more accurate representations of the data and increasingly less generalized portrayals of regional trends. A surface's "goodness of fit" is measured by R^2, the *coefficient of determination,* which ranges from 0 for the worst possible fit to 1.0 for a perfect fit, with all vertical deviations equal to zero. A surface with an R^2 of 0.5 will, for example, have a weaker, less well-defined trend than a surface of the same order of polynomial having an R^2 of 0.9 for another geographic distribution mapped for the same set of data points.

The coefficient of determination can be plotted against the order of polynomial for successively higher orders of trend surface in an effort to detect the most parsimonious representation of the regional trend. An elbow in this curve would indicate that further flexibility obtained by increasing the order of the trend surface is marginal (Fig. 3-6).

Although some applications benefit from mapping the general regional trend, other research questions might be answered more readily by a map of the distribution with the

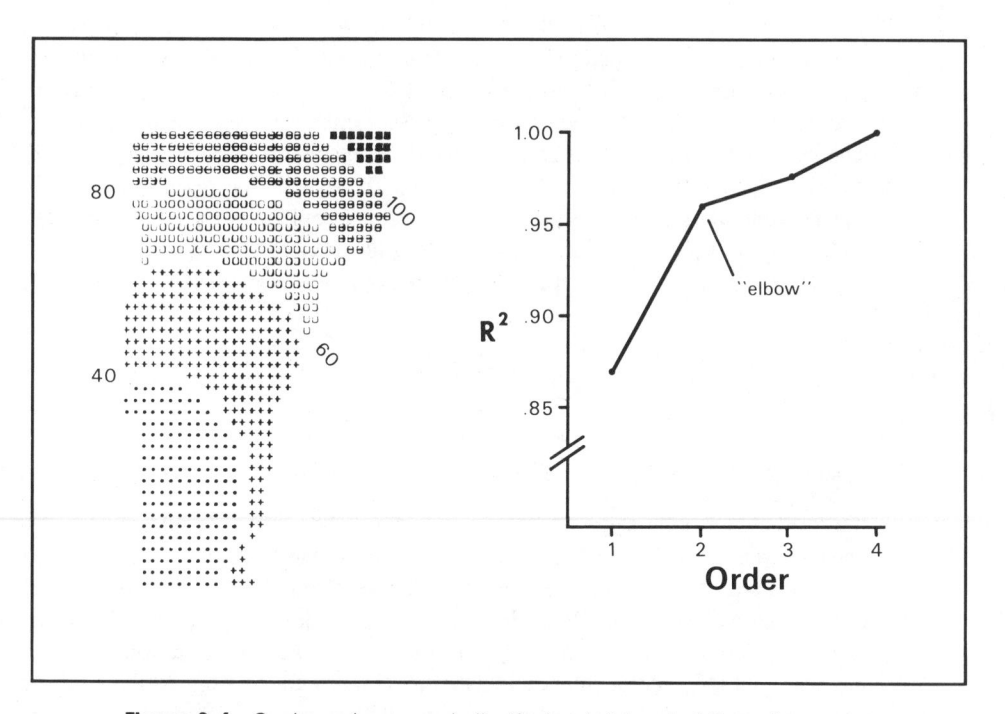

Figure 3-6 Contours show quadratic (2nd order) trend of Canadians as a percentage of foreign stock, for Vermont county center points. Plot of coefficient of determination (R^2) against surface order has an "elbow" at the 2nd order, after which higher order surfaces yield only marginal improvements in the "fit."

regional trend removed. A study of urban land values might, for example, require a test of the hypothesis that the price of a square meter of ground declines away from the city center. A quadratic surface can have a simple peak, which would be expected to occur over the commercial center of the city. A contour map of this surface might depict one or two major distortions of the declining-land-value model, such as the effect of a transportation axis, which might raise nearby land values.

The vertical deviations between the data points and the trend surface are called *residuals* and can be mapped to show the geographic pattern of departures from the general trend.[3] High, positive residuals would occur in areas with land values higher than might be expected from the overall quadratic trend, perhaps because of particularly desirable scenery. Low, negative residuals would occur where land values are less than the regional trend might suggest, perhaps depressed by proximity to the noxious odors of a chemical plant. In addition to the global solution provided by a trend-surface map, a complete cartographic treatment of a geographic phenomenon also requires contour maps of the original distribution and the trend residuals, obtained as local solutions, based only on nearby data points.

Packages

As with most sophisticated mapping systems, data for SYMAP are organized as modules called *packages*. These modules might be sections of a deck of punched cards or separate files stored in peripheral memory. The first three available packages contain coordinate information: an A-CONFORMOLINES package contains conformant zones, with the areal units for choropleth maps coded in vector form as area polygons, whereas an A-OUTLINES package indicates an irregularly shaped region within which contour symbolization is to appear and would be used with a B-DATA POINTS package containing the coordinates of data points for contour, proximal, and trend surface maps. The C-OTOLEGENDS package permits the printing within the rectangular neat line of north arrows, bar scales, and other legend features for which the user provides vector data. The D-BARRIERS package specifies impermeable barriers that prohibit data points on one side of a barrier from influencing values interpolated on the other side, as well as permeable barriers that merely diminish the influence of a data point on the opposite side. Barriers might be useful for showing the effects of unbridged rivers, lakes, swamps, and other obstacles on travel-time maps, for example, a map showing driving times to a city center from the outer suburbs.

Data values for the areal units and data points constitute the E-VALUES package. An E1-VALUES package can rearrange the order of these data values. The F-MAP package contains all other specifications for the map and is the only mandatory package: an F-MAP package is required for each map produced.

---

[3] Ronald Abler, John S. Adams, and Peter Gould, *Spatial Organization: The Geographer's View of the World* (Englewood Cliffs, N.J.: Prentice-Hall, Inc., 1971), pp. 124–37.

Some Electives in SYMAP

The F-MAP package usually consists of a series of numbered electives, of which 38 are available. The first two electives act together to determine the scale and areal coverage of the map. Elective 1 specifies the vertical and horizontal dimensions of the printed map, in inches. Elective 2 provides the coordinates, in whatever units the program user employed in the A and B packages, of the upper-left and lower-right corners of the rectangular window, or *viewport,* to be displayed within the map frame determined by elective 1 (Fig. 3-7). Electives 1 and 2 are useful for changing the scale of the map and permit the mapping of new windows as insets or different sheets in a series of maps covering different portions of a large region. Because the window will be stretched or compressed, as directed, to fit the neat line, poor coordination of these electives is likely to distort the map.

Instead of the standard *X-Y* axes directed to the right and upward, SYMAP uses coordinate axes extending downward and to the right of the origin. This reversed direction for the vertical axis is an artifact of an earlier version of the program in which all locations were specified as rows and columns on the printer grid, which is printed downward beginning with the first row at the top of the map. Data captured with a digitizer as standard *(x, y)* coordinates may be transformed to SYMAP coordinates by some other program. As an alternative, the origin of the digitizer can be set in the northwest corner, so that the *X* axis will yield the across coordinate and the *Y* axis, with the minus sign dropped, will yield the down coordinate.

Several electives establish mapping categories for the various graytone area symbols.

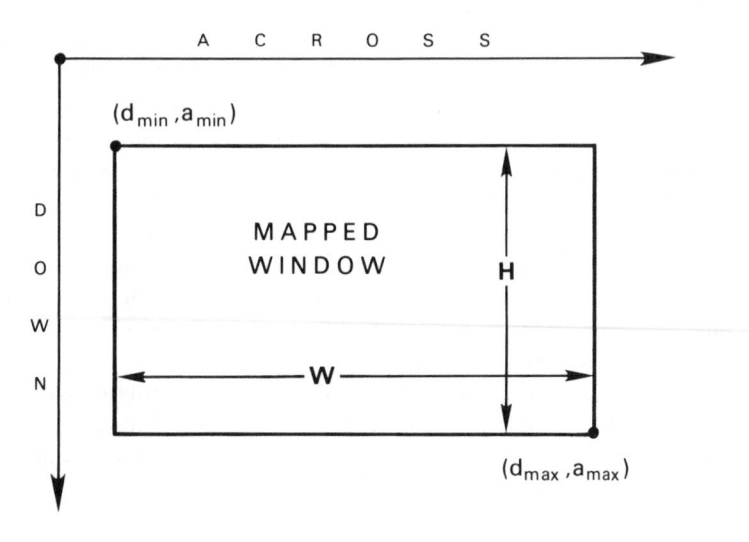

Figure 3-7 Areal extent of mapped window is given in SYMAP Elective 2 by coordinates of upper-left and lower-right corners. Size of map is specified in Elective 1 by height (H) and width (W), in inches, of the neat line.

Elective 3 sets the number of categories, which may range from 2 to 10. If this elective is not included in the F-MAP package, SYMAP will use five categories. Electives 4 and 5 set the minimum value for the lowest category and the maximum value for the highest category. The low and high values contained in the E-VALUES package will be used to establish the range of the classification if the user "defaults" by omitting these electives. Elective 6 apportions this range among the various categories according to relative weights specified by the program user. If this elective is not used, each class receives an equal portion of the classification range.

All four electives are usually needed to specify breaks between mapping categories. Suppose, for example, that there are to be four classes with class intervals 0.0 to 4.9, 5.0 to 9.9, 10.0 to 19.9, and 20.0 to 50.0. Elective 3 must invoke four categories, and electives 4 and 5 must specify a classification range from 0.0 to 50.0. Lower limits of 5.0, 10.0, and 20.0 result for classes 2, 3, and 4 if the respective weights given in elective 6 for the four categories are 5, 5, 10, and 30. Should any data values be negative or greater than 50, parts of the printer grid would be shaded with L's and H's, to indicate values lower and higher than the specified range.

Quantile (equal rank) categories, with approximately the same number of data values in each category, can be useful if the map viewer might need to focus upon, say, the highest fifth or lowest fifth of the areal units on a choropleth map. Setting the weights for all categories in elective 6 to zero directs SYMAP to sort the data values from lowest to highest and assign categories on the basis of rank.

Other programs have been developed, particularly by geologists, for displaying trend surfaces and contour maps with a line printer, but SYMAP is likely to be more useful because of the comparative ease with which the program user can control the appearance and information content of the map. The most visually effective SYMAP can be selected from numerous views of the same distribution obtained with numerous F-MAP packages specifying various trial classifications. The program might be made even more flexible, of course, if the user were able, for example, to request all quantile and equal-interval maps with four, five, and six categories in a single F-MAP package. Better still would be the incorporation of an optimization algorithm testing numerous sets of class breaks and selecting for mapping that classification with the "most natural," and most accurate, natural breaks. The program thus might be instructed by its user to screen without printing an indefinitely large number of maps that might be produced.[4] Each mappable pattern would receive, for example, an accuracy score, and only the most accurate map would be printed.

The information content of a SYMAP might be determined by the classification imposed upon the data, but this information is unlikely to be conveyed adequately to the map reader if the graphic symbols produced by the line printer are poorly differentiated. Elective 7 changes the area shadings and other symbols printed within the neat line. A

---

[4]George F. Jenks and Fred C. Caspall, "Error on Choroplethic Maps: Definition, Measurement, Reduction," *Annals of the Association of American Geographers,* 61, no. 2 (June 1971), 217–44.

user not satisfied with the gray scale produced by printing a pattern of bold periods for the lowest class and a pattern of faint minus signs for a slightly higher class thus can adopt a new set of area symbols adapted to the print chain and impact setting of the line printer. Different sets of shading symbols might be appropriate: one for fresh, dark ribbons, another for faint, worn ribbons.

  Embellishments such as large type for a title and major highways to provide a geographic frame of reference are best added manually after the map is printed. Nonetheless, the C-OTOLEGENDS package enables the user to specify a more convenient legend within the map border. SYMAP prints a legend below the neat line in a position for the map key seldom used on hand-drawn maps. The printed size of this standard SYMAP legend also cannot be enlarged or reduced by the user in anticipation of a change in the scale of the map, and reflects the difficulty in writing a computer program to accommodate a variety of unpredictable user needs. Yet with the C-OTOLEGENDS option the legend not only can be positioned appropriately on the map but also can be enlarged or reduced in coordination with the map as a whole. These cosmetic additions can be helpful if SYMAP is to produce large-scale wall maps, perhaps for an oral presentation before a small group of planning board members, administrators, or managers.

  SYMAP demonstrates the need for temporary data files, also called *scratch files,* to store results generated as an intermediate step between the data entered in the various packages and the final line-printer map. A scratch file represents each row and column of the final map in a single rectangular array of surface elevations, instead of the separate strips generally no wider than the 130 columns used to accommodate wide maps on the printer. Scratch files are also used to provide storage in external memory for the areal units represented in the A-CONFORMOLINES package, the regional outline from the A-OUTLINES package, and cosmetic adornments from the C-OTOLEGENDS package.

  When SYMAP prints the map in strips, each row of each strip is read from the temporary file containing the interpolated elevation grid. For each grid cell the surface elevation is compared with the class intervals so that the appropriate overprint characters can be transferred to the corresponding column of the four-row array used for composing an individual line of the strip. Legend characters retrieved from another scratch file replace any shading symbols competing for the same positions on the printer grid. Area shading is also suppressed on trend surface, proximal, and standard contour maps for cells lying outside the regional outline. These scratch files are saved during the execution of the program so that lengthy computations need not be repeated for subsequent maps if no changes have been made. For some applications, computational efficiency would be improved even further if scratch files requiring no alterations could be saved between separate executions of the program.

Grid Interpolation

In addition to estimating the elevations of the surface at locations without data points, SYMAP interpolation must relate the plane coordinates provided by the user to positions on the line-printer grid. Cells of the printer grid may be 0.25 by 0.25 cm (0.1 by 0.1

in.), but more commonly the vertical dimension exceeds the horizontal dimension to accommodate the familiar vertical elongation of letters and numbers. SYMAP can adapt easily to the rectangular form of the grid cells of most line printers because the user can specify, in elective 15, the numbers of rows n_r and columns n_c per inch. The formulas

$$r = 1 + \text{int}\left(\frac{d}{d_{\max} - d_{\min}} H n_r\right)$$

and

$$c = 1 + \text{int}\left(\frac{a}{a_{\max} - a_{\min}} W n_c\right)$$

can then convert down and across coordinates *(d, a)* to the corresponding grid-cell coordinates *(r, c)*. Appropriate scaling is specified by the height H and width W of the neat line, given in inches in elective 1, and by the coordinates of the upper-left $(d_{\min}, a_{\min})$ and lower-right $(d_{\max}, a_{\max})$ corners of the neat line, given in the same units as d and a in elective 2 and the A and B packages (Fig. 3-7). The *floor function* int truncates the expression within the outer parentheses to the closest lower integer.

Because the elevation grid is processed systematically, row by row and within a row, column by column, inverse formulas are needed for converting from *(r, c)* to the *(d, a)* coordinates used during interpolation for computing distances between control points and the centers of the grid cells. The equations

$$d = [0.5 + (r - 1)] \, (d_{\max} - d_{\min}) \frac{H}{n_r}$$

and

$$a = [0.5 + (c - 1)] \, (a_{\max} - a_{\min}) \frac{W}{n_c}$$

provide the necessary transformations.

A two-stage procedure conserves the central processor time required for interpolation while maintaining graphic accuracy. The first stage uses a relatively complicated and precise algorithm to estimate elevations for the centers of grid cells spaced two rows and three columns apart. In the second stage, simple linear interpolation estimates surface elevations for the remaining five-sixths of the cells on the finer grid of the line-printer map. Although the user can elect, in the F-MAP package, to have the more precise algorithm estimate surface values at all cells of the map, the visual coarseness of the line-printer map ordinarily does not warrant the substantial increase in CPU time needed to effect only minor changes in the look of the map.

Linear interpolation distributes the elevations interpolated for the coarser, first-stage grid by computing simple weighted averages for intermediate cells (Fig. 3-8). For fine-

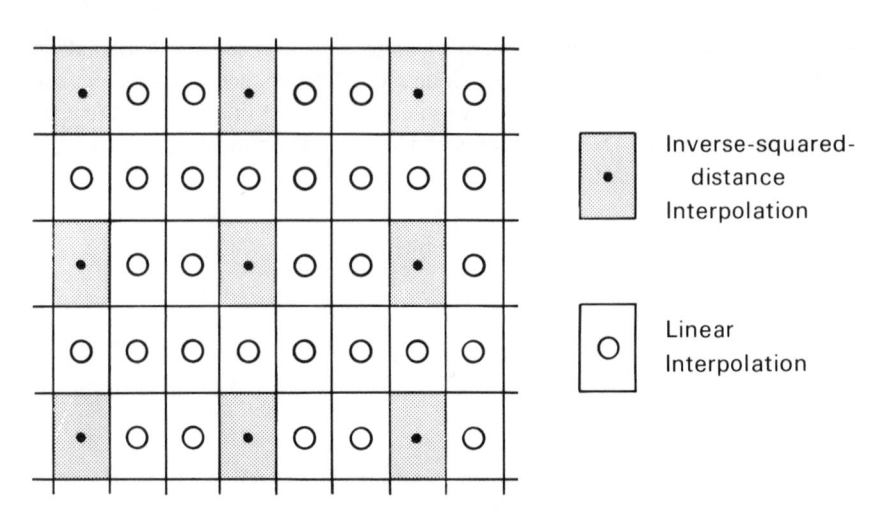

Figure 3-8 Inverse-squared-distance interpolation from the data points estimates the elevations of points centered within the cells of a coarse grid. Linear interpolation extends these more accurately estimated elevations to the intermediate cells of a finer grid.

grid cells sharing a column with coarse-grid cells, surface elevation can be estimated by merely adding the elevations at the cells above and below and then dividing by 2. For intermediate cells sharing a row with interpolated cells, elevation estimates require inverse weighting according to distance from a coarse-grid cell: the intermediate value is approximated by taking two-thirds of the elevation of the closer of the two nearby interpolated cells in the row and adding one-third of the elevation of the other interpolated cell. Elevations for the remaining intermediate cells not aligned in a row or column with an interpolated cell are estimated by a slightly more complicated averaging formula,

$$z_{\text{intermediate}} = \frac{\sum\limits_{i\,=\,1}^{4} (1/d_i)z_i}{\sum\limits_{i\,=\,1}^{4} (1/d_i)}$$

which uses as weights the inverses of the actual distances d_i between the center of the intermediate, fine-grid cell and the centers of each of the four closest interpolated cells with interpolated elevations z_i. These distances can be computed according to the equation

$$d_{pi} = [(x_i - x_p)^2 + (y_i - y_p)^2]^{1/2}$$

for the straight-line distance d_{pi} between points i and p.

The more precise, first stage of SYMAP interpolation reduces the contributions to estimated elevations of relatively distant control points. In the more general formula for a weighted average computed from n known elevations,

$$z_{\text{estimated}} = \frac{\displaystyle\sum_{i=1}^{n} w_i z_i}{\displaystyle\sum_{i=1}^{n} w_i}$$

simple linear interpolation uses weights w_i of $(1/d_i)$, whereas first-stage SYMAP interpolation uses weights of $(1/d_i^2)$. This inverse-square weighting diminishes the influence of distant data points in the same way that the influences of gravity and electrical attraction decline as the inverse of the square of distance.

Interpolation is a highly subjective process, and an estimation procedure is not right or wrong, but merely plausible or absurd. The inverse-squared-distance weighting used by SYMAP is a reasonable approach, grounded in the classical mechanics of Newton and Gauss, but surely not the only plausible approach. For example, the closest control point might exert a much greater influence on interpolated elevations based upon the formula

$$z_p = \frac{1}{2} z_{\text{closest}} + \frac{1}{2} \left[\frac{\displaystyle\sum_{i=1}^{n} (1/d_i)z_i}{\displaystyle\sum_{i=1}^{n} (1/d_i)} \right]$$

which estimates z_p as the average of the closest elevation and the result given by simple linear interpolation. If the surface is well behaved, more smooth than rough, and with no marked irregularities, the closest known elevation should approximate better, albeit not precisely, an unknown elevation than the height of the surface at any other point.[5]

Because distant data points have little influence on elevations estimated with inverse-squared-distance weighting, further reductions in processing time result from basing estimates on only the closer control points. For SYMAP, "closer" is defined by a circular *search region* centered upon the interpolated point, the center of a cell in the coarse grid. Unless otherwise directed by F-MAP electives, SYMAP uses a search region containing between four and ten data points. Each search begins with a region defined by initial search radius r_{init}, computed as

$$r_{\text{init}} = \left(\frac{7A}{\pi N} \right)^{1/2}$$

[5]Waldo R. Tobler, ed., *Selected Computer Programs* (Ann Arbor: Department of Geography, University of Michigan, 1970), p. 55.

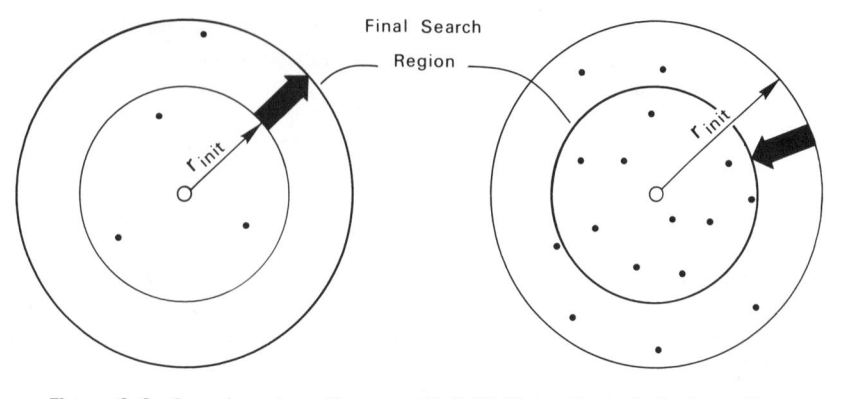

Figure 3-9 Search region will expand (left) if initial radius includes fewer than 4 points or contract (right) if initial radius includes more than 10 points.

which would in principle include an average of seven control points in the search region if the N control points in the B-DATA POINTS package were distributed over a mapped region with area A according to a hexagonal, closest-packing arrangement.

The search begins by computing the distances from interpolated point p to all data points. If fewer than four control points fall within r_{init} of point p, the search region expands until the radius includes at least four points (Fig. 3-9). If more than ten points are included, the search region shrinks to contain only the ten points closest to the interpolated point. The final search region thus reflects local variations in the spacing of control points. But each time the program computes an elevation for a coarse-grid point, it must examine all known data points to find out which ones are the closest. Thus, the interpolation process may be very time consuming if the number of data points is large. Unlike a human, the computer cannot "see" the circles in Figure 3-9.

Through F-MAP electives, the user can specify the minimum and maximum numbers of data points for interpolation, a constant initial search radius, and a maximum search radius. These interpolation parameters permit the map author to alter the form of the interpolated surface. An increase in any of these specifications tends to increase the size of the average search region, thereby basing elevation estimates upon more data values and producing a smoother, less jagged surface with less irregular, more gently flowing contours. If the control points are highly clustered in a small portion of the region, an increase in the minimum number of points for interpolation is likely to affect most individual estimates and, in areas with few data points, the entire surface as well. The user must be aware of these possibilities and determine through experimentation that the particular contoured SYMAP is a portrayal of the surface appropriate to the purpose of the map.[6]

Control points equidistant from the center of the search region need not be equally

---

[6]Elri Liebenberg, "Symap: Its Uses and Abuses," *Cartographic Journal,* 13, no. 1 (June 1976), pp. 26–36.

important: two data points in the same general direction away from the interpolated point are somewhat redundant, whereas two points diametrically opposite each other provide more information about surface elevations. Figure 3-10 illustrates a search region with seven data points: a single point isolated from a cluster of six other points. Consider only the vertical profile through the interpolated surface between the isolated point and the cluster. If all seven points were weighted equally in estimating the elevation midway between the single point and the cluster, the interpolated value would be biased quite heavily toward the average elevation of the clustered points, and the profile would be curved rather than straight.

SYMAP adjusts the weights of the data points according to each point's *directional isolation* from the other control points within the search region. The weight w_i, assigned data point i according to the inverse of the squared distance from the interpolation point p, is reduced for directionally clustered points and increased for directionally isolated points. A point's directional isolation is measured by considering the cosine of the angle $A(i, p, j)$ between a line to the center p of the search region and a line to the center from another control point j. For diametrically opposite points the angle is 180 degrees, the cosine is -1, and the expression $(1 - \cos A(i, p, j))$ is 2, whereas for two points i and j lying along the same radial line from the center the angle is 0 degrees, the cosine is 1, and $(1 - \cos A(i, p, j))$ yields 0. The relative directional isolation of a control point, determined by summing the values of this expression for directional angles to all other data points within the search region, is used to modify the weights so that, in the example considered in Figure 3-10, a more plausible straight slope would result.

A further modification counteracts the undesirable tendency of inverse-squared-distance weighting to produce horizontal surfaces in the vicinity of control points. As distance between an interpolated point and data point i diminishes, the inverse of the squared distance $(1/d_{ip})^2$ increases, producing an extremely large relative weight that can negate

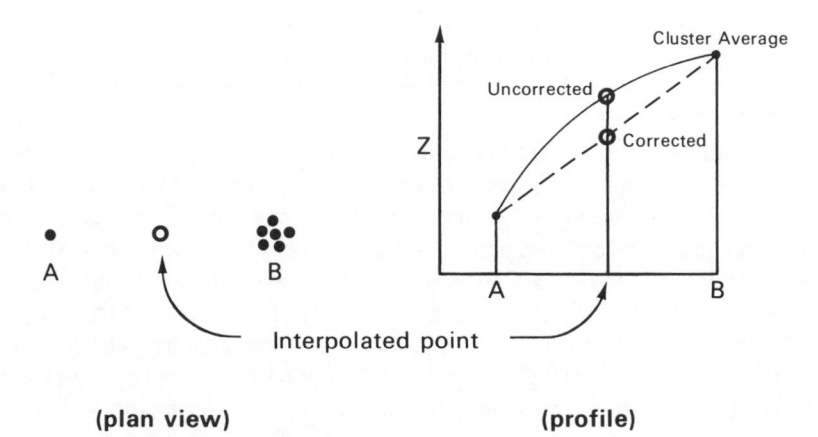

Figure 3-10 Without correction for directional isolation, interpolated elevations would be unduly biased toward generally similar elevations at a cluster of data points *(B)*.

the influence of all other control points in the search region. Although this effect is appropriate because the surface at a control point obviously should have the elevation of the control point, a small, unrealistically flat plateau occurs around each data point. These unwarranted horizontal terraces can be prevented by projecting slope as well as elevation across space. SYMAP thus computes for all data points a weighted average of the slopes to other nearby data points. Where an interpolated point is relatively close to a control point, the control point's average slope assumes a greater role than its elevation in the estimation process.[7]

This interpolation of slopes as well as elevations might produce unexpected results for program users unaware of its operation. Local high and low points on the interpolated surface commonly occur at locations other than data points. Elevations measured only around the sides of an embankment encircling a low plateau, for example, might be extrapolated upward by SYMAP to form a rounded, higher dome. Unless directed otherwise by electives controlling the minimum and maximum interpolated values, SYMAP will not extrapolate to values lower than the lowest or higher than the highest data values provided by the user. Moreover, unless the elective for fractional extrapolation is invoked, an extrapolated elevation will not fall below the lowest known elevation within the search region by more than 10 percent of the range of data values. A similar restriction also holds for extrapolation upward, beyond a locally high elevation. With climatic data, for example, the user might set the extrapolation electives to permit estimation beyond the absolute and local extrema, whereas with some sociocultural maps, particularly with percentages that may not logically be negative or exceed 100, definite limits might be mandated by the theme of the map.

The fringes of the contoured region are less likely to be portrayed accurately than the interior, because slopes can seldom be projected substantial distances beyond the area of known elevations. The user requiring a more accurate contoured surface should include in the data representative control points immediately outside the perimeter of the mapped region. Interpolation thus replaces the less accurate extrapolation in fringe areas. This adjustment is valid, of course, only if the processes responsible for the mapped pattern are similar both inside and outside the mapped region; a contoured map of median family incomes for Texas should be improved, for example, by considering observation points in nearby Oklahoma, but not across the Rio Grande in Mexico.

The locations of the control points within the mapped region are also important. The user able to control the geometry of the sample might produce an accurate map by specifying a uniform sampling grid with the density of the control points adjusted to the variability of the mapped phenomenon. Experimentation with sampling densities might be warranted in order to ascertain the stability of the mapped pattern. For *isopleth* maps, with a surface estimated from control points within areal units such as census tracts and counties, supplemental control points might prove useful for large or exceptionally stringy areas. Although it is customary to locate data points for interpolation at the approximate

---

[7]Donald Shepard, "A Two-Dimensional Interpolation Function for Irregularly-Spaced Data," *Proceedings*, Association for Computing Machinery, 23rd National Conference, 1968, pp. 517–24.

4

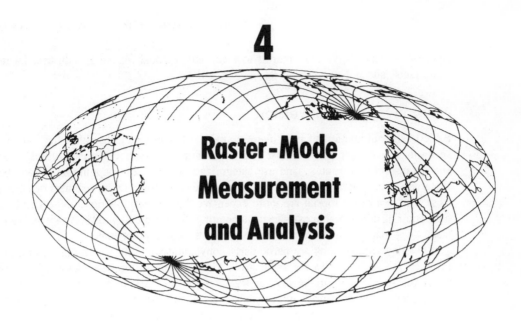

Raster-Mode Measurement and Analysis

Raster information can be displayed either directly or after an analysis designed to extract from the data information more useful than commonly found on a single printed map. This chapter discusses procedures for analyzing multiple distributions, as overlays in a land-information system or as data from multispectral scanners mounted on a high-altitude aircraft or orbiting satellite. Also considered are measurements that can be made from an elevation surface, or digital terrain model, to yield slope maps and shaded relief maps.

OVERLAY ANALYSIS

Map users vary greatly in their needs. Many users are concerned in some way with comparing geographic distributions by comparing maps. These cross-comparisons might lead to estimates of the similarity, or cross-correlation, of two distributions, or to evaluations of potential sites for a new factory or health-care facility based upon maps of various relevant factors. Proposed land uses can be favored by proximity to markets and raw materials, inexpensive land, a competent and willing labor supply, and good transportation, and repelled by expensive site development costs, restrictive zoning, and high transportation costs. In some cases the map user might attempt to superimpose mentally two or more maps. If a more precise solution is needed, the map user might draw two or more distributions on the same sheet of paper and symbolize this composite map to reveal areas that are suitable, only marginally suitable, or entirely unsuitable. Computer programs using files of raster data can rapidly and accurately perform many different types of overlay analysis of interest to local and regional planners, industrial location consultants, and military strategists.[1]

Polygon-to-grid Conversion

Basic to the automation of map overlay analysis has been an algorithm for converting vector data to a raster format. Most data representing zoning boundaries, political boundaries, transportation routes, land-use categorizations, and soil-mapping units have been captured as vector data by manually digitizing area polygons. Drum scanners that capture these data in a raster format are more expensive than line-following digitizers and yield files requiring considerable editing and annotating. Nonetheless, good software and careful manuscript map preparation have made optical scanning practicable.

Overlay comparisons are particularly convenient when all geographic distributions involved are registered to the same grid. Although the raster data files need not even reside on the same magnetic disk or reel of tape, the various overlays can be viewed conceptually as a set of grids neatly stacked so that all data for a rectangular plot of land can be retrieved by specifying the same row and column coordinates (Fig. 4-1). The complicated, expensive, and often redundant computing required to determine the intersections of area polygons argues persuasively against an analysis system operating directly with vector data. The availability in raster format of land-use and land-cover data captured with orbiting satellites is an additional incentive for raster-mode overlay analysis.

Queries to a digital land-analysis system might specify that analysis be restricted to a specific geographic zone such as a town, city, or special taxing district. Each cell within

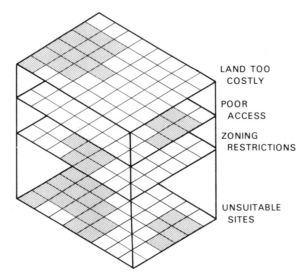

LAND TOO
  COSTLY

POOR
  ACCESS

ZONING
  RESTRICTIONS

UNSUITABLE
SITES

Figure 4-1 Raster-mode overlay analysis, with grids superimposed to determine sites unsuitable because of land cost, poor access, or zoning restrictions.

---

[1]Devon M. Schneider with Syed Amanullah, *Computer-Assisted Land Resources Planning* (Chicago: American Planning Association, Planning Advisory Service, Report No. 339, 1979).

this territory must be tagged appropriately in the grid overlay representing that level of administrative unit. The polygon-to-grid conversion required for identifying the zone's cells has two principal stages: transferring the linear boundaries of the polygon to the grid and filling in the interior cells.

Transfer to a grid of the linear edges of a polygon is straightforward: after polygon vertexes are identified in grid cells, these vertex cells and all intervening cells are filled with a number called the *parity,* indicating the "equality" of cells representing the same boundary. The transformation from standard plane coordinates (x, y) to grid coordinates (I, J) is given by the formulas

$$I = 1 + \text{int} \left(\frac{y - y_{\text{top}}}{DY} \right)$$

and

$$J = 1 + \text{int} \left(\frac{x}{DX} \right)$$

for a grid with cells DX wide and DY high and with rows numbered from the top downward (Fig. 4-2). The rows of the grid parallel the X axis, and the columns parallel the Y axis. The plane coordinates of the upper-left corner of the cell in the first column of the first row are $(0, y_{\text{top}})$. Floor function int truncates decimal numbers to integers referencing rows and columns. If a map is to be displayed on a standard line printer, the ratio of DX to DY can accommodate the elongation of the line-printer grid cells.

Cells between the vertex cells of a line segment bounding a polygon can be identified by computing distances from the centers of intervening cells to a straight line connecting the actual vertex points (Fig. 4-3). If the vector-to-raster conversion is intended ultimately to represent a linear feature such as a river, all rows and columns between the two vertex cells must contain at least one cell with the parity representing the feature; otherwise, the raster image of the linear feature will be broken rather than continuous. If more rows than columns intervene, stepwise extension of the line progresses row by row. After a cell is added to the line, the next cell in the same column but in the next row down the feature is compared with its two neighboring cells in the new row. For each of these three cells, distance is computed from the cell center to the point where the feature intersects a straight line through the middle of the row. The cell with the shortest distance, that is, the cell lying closest to the linear feature, receives the parity identifying the feature. This process is replicated column by column, instead of row by row, if more columns than rows intervene between vertex cells.

Chains of grid cells representing boundaries and linear features are sometimes stored in data bases according to a simple, three-bit code that points to the next cell in the chain with an integer in the range from 0 to 7. A simple counterclockwise convention assigns these eight integers to the eight cells adjoining at the sides and corners of the cell in

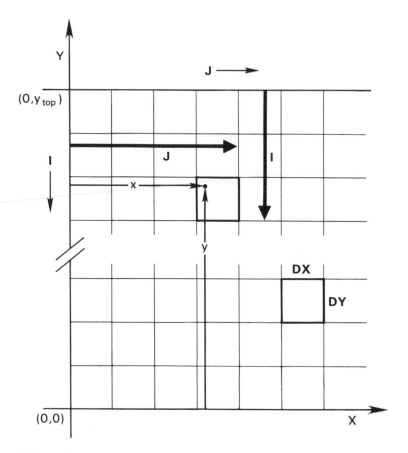

Figure 4-2 Plane coordinates *(x, y)* can be transformed to grid cell coordinates *(I, J)*.

question. The cell immediately to the right is designated by the code 0, the cell diagonally to the upper right by the code 1, the cell directly above by the code 2, and so forth. A sequential list of these direction codes, called *Freeman chain encoding* after its originator, is a convenient method for conserving memory by *packing* two grid coordinates *I* and *J* into a single two-byte word.

Vectors describing area polygons normally are coded in a consistent, clockwise direction that will place the interior of the area to the right of the boundary. For areal units with holes called *enclaves* (an example often cited is the city of Hamtramck, Michigan, independent from but completely surrounded by Detroit), a cut-line extending inward from the exterior boundary can be retraced outward after the interior boundary encircles the enclave in the counterclockwise direction (Fig. 4-4). The clockwise convention places to the left territory not belonging to the area of interest. All traces of the

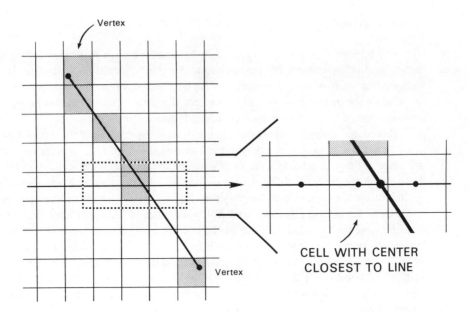

Vertex

Vertex

CELL WITH CENTER
CLOSEST TO LINE

Figure 4-3 Line segment is completed between vertex cells by including in each row the cell with its center closest to the line.

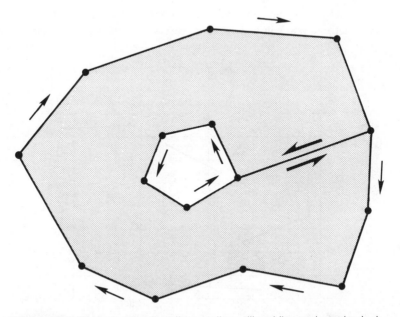

Figure 4-4 Clockwise directional convention, with cut-line and counterclockwise direction shown for enclave. Area of interest always lies to the right.

imaginary boundary between the area and itself formed by the cut-line are removed after processing.

The interior of a polygon is filled in by scanning each row from left to right and setting grid cells between the boundaries to the parity identifying the area. In general, parity filling starts when a boundary cell is encountered during the scan across a row and stops when the next boundary cell is reached. If a third boundary cell is detected, filling resumes until a fourth boundary cell is found, and so forth.

Directional coding enables the computer to interpret accurately the gridded boundaries. When the boundary of a polygon is transferred to a grid, the directions of the edges are noted in the grid cells. For the clockwise convention, boundary cells might be marked L, to designate the left side of the areal unit, for edge vectors with an upward trend, or R, to indicate the right side of the polygon, for edge vectors with a downward trend (Fig. 4-5). Cells having no effect on area filling are marked N, for neutral. For example, a boundary reentering a cell, as for the cell at the neck of a polygon in the form of an hourglass, resets the directional indicator to N. An L cell encountered on a left-to-right scan initiates the filling-in of all subsequent cells until an R cell is detected. An N cell is filled with the parity code for the areal unit without starting or stopping the filling-in process. Figure 4-5 demonstrates more fully the directional coding used for this polygon-to-grid algorithm.

In grid-based information systems, used widely in Sweden and other European countries for regional planning and census mapping, individual cells are commonly assigned to a single administrative or political unit of a particular class, even though a substantial part of the cell might fall within some other jurisdiction. When a boundary cell of an area polygon is needed to preserve the connectivity of the boundary, that cell

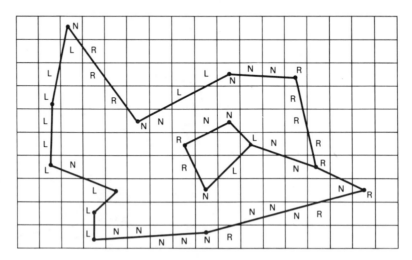

Figure 4-5 Directional codes indicate left-hand (L), right-hand (R) and neutral (N) cells. Scanning across rows from left to right will fill in cells between a matching pair of L and R cells with parity identifying the area of interest.

ought not to be assigned to the area if another area polygon covers a greater portion of the land represented by the cell. An appropriate adjustment can be made after polygon-to-grid conversion if the percentage of each boundary cell lying within the polygon is recorded during the conversion process. A grid cell can then be assigned uniquely to the polygon covering more of the cell than any other polygon.

Areal Aggregation

Digital geographic data bases stored in raster mode permit many different types of analysis. In a sense, a geographic base file is a form of map that can be read by the computer and used either to compile other maps to be read by humans or to generate purely numerical results. The U.S. Geological Survey, for example, uses a raster file with square cells representing land units approximately 200 m (656 ft) on a side, and thus containing 4 hectares (about 10 acres), for the storage and analysis of land-use and land-cover information. Overlays are included for states and counties, census tracts within metropolitan areas, townships, and other minor civil divisions, hydrologic units such as individual watersheds, federally owned land, and state-owned land.

In addition to a variety of graphic displays, the U.S. Geological Survey system, called GIRAS for Geographic Information Retrieval and Analysis System, generates land-use summary statistics for counties and census tracts.[2] These summaries estimate, for example, the percentage of a county covered by strip mines, quarries, and gravel pits or the percentage of a census tract occupied by land devoted to commercial and service activities. Planners and environmental researchers can use these statistical summaries for comparisons with the social, economic, and cultural data available from the Bureau of the Census and with other governmental records. Relationships that might be explored include the influences of population growth on the decline in agricultural land. This data-management system also permits the systematic year-to-year monitoring of land-use change, an important indicator of both environmental quality and the nation's capacity to feed itself and others.

Set Theory and Land Analysis

Areal summaries based upon administrative and political units illustrate the ease with which raster data can be manipulated by the basic operators of set theory, union and intersection. The area of a county's cropland, for instance, is merely the area of the intersection of the county and all cropland: cells falling within the county on the political overlay and designated as cropland on the land-use map need only be totaled and multiplied by the unit area for a cell. These county-cropland cells can be displayed graphically as well as reported numerically as a sum. The union operator permits two or more land

---

[2]William B. Mitchell and others, *GIRAS: A Geographic Information Retrieval and Analysis System for Handling Land Use and Land Cover Data* (Washington, D.C.: U.S. Government Printing Office, Geological Survey Professional Paper No. 1059, 1977).

categories to be combined, as for example, the union of cropland and pasture-grassland would illustrate the distribution of nonforested agricultural land.

Composite maps can portray the suitability of land for a specific administrative action or a zoning classification. Each land characteristic, such as slope, soil permeability, vegetative cover, and present use, can be assigned a weight. These weights are combined in a weighting function used to compute for each cell a composite index serving as a summary evaluation of the parcel's merits and liabilities as, for example, the site of a new regional shopping center.

Factors completely limiting development can, of course, be used to exclude cells from consideration. A cell might be excluded because of a single undesirable property; for example, swampland not only is a costly construction site but might be protected by law as a wetlands habitat. In this example the set theoretic difference operator would be used to subtract from the set of otherwise acceptable cells all cells belonging to the set of wetland cells. If a composite score below some known threshold is sufficient to exclude a cell from further consideration, this threshold and the weighting function can be used to define a new set of low-scoring cells, to be subtracted from the solution set. Negative factors might be segregated from positive factors and combined with appropriate weights in a penalty function; all cells with a penalty index above a stated threshold would then be removed from the solution set.

More than one penalty function or excluding property can be used. A cell thus might be ignored as a potential site for recreational development because, for example, either (1) the potential soil erosion exceeds 2,000 kg/hectare/year (about 0.89 ton/acre/year) or (2) the soil is poorly drained and has an erosion potential greater than 500 kg/hectare/year. Several criteria can be combined with OR (union) as well as AND (intersection) logical operators to integrate many different land attributes in an evaluation of development limitations and opportunities.

Binary arithmetic provides a convenient method for performing set theoretic operations. Each set can be represented by a grid of binary numbers, called a *bit plane;* a cell belonging to the set is coded 1, and a cell not belonging to the set is coded 0. Consider two cells belonging to sets A and B. The intersection of the sets results from multiplying the binary digits, so that $A \cap B$ results from $A \cdot B$. That is, a cell is not included in the intersection unless both A and B are coded 1, the only situation for which the product will equal 1 instead of 0. The union can be expressed as the complement of the intersection of the sets A^c and B^c complementing sets A and B. Thus, $A \cup B$ is represented by $(A^c \cap B^c)^c$, where, for example, A^c is the set of all cells not in set A. The complement of a set is simply the difference between the universe W of all cells and the set in question. In binary arithmetic, set A^c can be derived from set A as $(1 - A)$ and the union $A \cup B$ can be computed as

$$A \cup B = (1 - ((1 - A) + (1 - B)))$$

which yields 1 only if A and B are both 1.

Bit-plane coding affords a convenient method for manipulating a large amount of data in internal memory, thereby saving the execution time and cost needed for retrieving data from external data files. Instead of storing only one 400 by 400 grid in a unique two-dimensional array, a computer with a 32-bit word can store 32 overlay grids in a single 400 by 400 array. Subroutines that can be called from programs written in FOR-TRAN and other high-level languages permit the packing of information for several grid cells in each word of computer memory. For efficiency, software subroutines for packing, unpacking, and manipulating bit planes might be written in assembly language, so that cell values can be computed according to the rules of Boolean algebra, whereby the laws $0 \cdot 0 = 1$ and $1 + 1 = 1$ for multiplication and addition yield the binary results appropriate for intersection and union.[3] These functions are provided even more conveniently and efficiently by specially designed computers called *array processors*.

Hash coding is another method for conserving the memory required for storing grid-cell information. Frequently, the area of interest is irregular rather than rectangular in form. A classic case in point is New York State, which would waste much money on empty cells if a single rectangular grid were to include those parts of the Northeast and adjacent Canada needed to accommodate the state's appendages to the east and north. The now defunct Land Use and Natural Resource (LUNR) inventory, developed when state-coordinated land-use planning was believed politically possible, used *hashing functions* to relate position in a more fully utilized data array to the UTM (Universal Transverse Mercator) grid coordinates found on all modern, large-scale topographic maps. The state was covered with four rectangular belts in order to reduce the number of wasted, empty cells.[4] Each square-kilometer cell was referenced by the UTM coordinates of its southwest corner, and each rectangular zone had a separate hashing function.

Figure 4-6 demonstrates the principles of hash coding. Three smaller grids with a

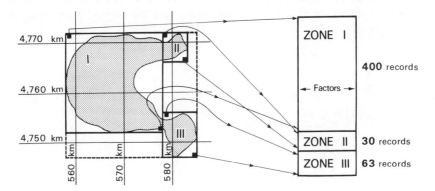

Figure 4-6 Hashing functions relate UTM coordinates *(N, E)* to records in data array (right) for three rectangular zones (left) covering inventoried region. Arrows indicate cell-record linkages for grid stored row by row in the data array.

---

[3]Niklaus Wirth, *Algorithm + Data Structures = Programs* (Englewood Cliffs, N.J.: Prentice-Hall, Inc., 1976), pp. 29–34.

[4]Robert Crowder, *LUNR, Land Use and National Resources Inventory: What It Is and How It Is Used* (Albany: New York Office of Planning Services, 1972).

total of 493 cells have replaced a larger grid with 675 cells. The cells are stored row by row in separate groups, for each of the three zones, as records in a large data array, the columns of which represent various land-use categories. Data are stored and retrieved by determining first the zone containing the grid cell in question. A cell's position within its grid is then converted to its sequential, row-by-row, cell-after-cell "count" from the cell in the extreme upper left, for example, 41 for the first cell in the third row of a grid with 20 columns. For cells in zone I, the count is also the record number in the large array. For cells in zone II, 400, the number of records in zone I, must be added to the count, and for zone III, 430 must be added.

Local Operators

Gridded maps can be used to compute other gridded maps reflecting a higher level of measurement or generalization. For example, an elevation grid, one form of a *digital terrain model,* can serve as data for generating a slope map useful for analysis of agricultural land capability. Elevation maps also provide data for maps of aspect, the direction of steepest slope, useful in studies of the microclimatic effects of differential warming and the angle of incidence of solar radiation. Digital elevation models (DEMs) can also provide the additional information needed for the automatic interpolation of meteorological data, which frequently must be corrected for elevation above sea level to provide an interpretable picture of atmospheric conditions. Highway engineers can use digital elevation models to estimate the amounts of surface material that must either be removed from a cut or used to fill a depression. When used in combination with gridded land-use information, digital elevation data can guide a computer·in the preparation of a dot-distribution map by indicating positions where the mapped phenomena are either likely or not likely to occur.

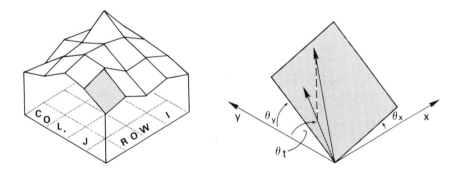

Figure 4-7   Angle θ_t indicates the direction of maximum slope for a quadrilateral above a cell of the elevation grid (left). Tilt of the quadrilateral (right) can be measured as the tangents of angles θ_x and θ_y.

The smoothing function mentioned earlier provided one illustration of a local operator (Fig. 3-11). Procedures for computing maximum slope and aspect also are local operators, because measurements of gradient and slope direction at a cell must be based upon elevations at neighboring cells. If a cell in an elevation grid is considered a tilted quadrilateral in three-dimensional space, the form of the surface above the cell is indicated by angles θ_x and θ_y measuring inclinations at the lower-left corner to the right along the row and toward the top along the column, respectively (Fig. 4-7). The slopes in the X and Y directions are given by the tangents $\tan \theta_x$ and $\tan \theta_y$, which are combined in the formula

$$\nabla(X, Y) = \left[(\tan \theta_x)^2 + (\tan \theta_y)^2 \right]^{1/2}$$

for the *gradient* $\nabla(X, Y)$. The direction of this maximum slope, measured as angle θ_t, clockwise from the positive Y axis (upward along the columns), can be estimated by finding the inverse angle θ_t of the trigonometric functions

$$\cos \theta_t = \frac{\tan \theta_y}{\nabla(X, Y)}$$

and

$$\sin \theta_t = \frac{\tan \theta_x}{\nabla(X, Y)}$$

The tangents of inclination angles θ_x and θ_y must be estimated from the elevations of other cells within a local search region centered separately on each cell in the grid.

Several methods can be used to estimate the directional slope angles for a simple 3 by 3 subgrid centered on cell (I, J). The simplest method considers only single pairs of cells for the row and column intersecting at the cell in question. The tangent along row I would be computed as

$$\tan \theta_x = \frac{E(I, J + 1) - E(I, J - 1)}{2g}$$

the difference in elevation divided by twice the ground distance g between centers of adjoining cells. For the tangent along column J, the equivalent formula is

$$\tan \theta_y = \frac{E(I - 1, J) - E(I + 1, J)}{2g}$$

More elaborate estimates of slope might fit a linear trend surface to all nine elevations in the 3 by 3 local subgrid and use as directional tangents the slope coefficients in the polynomial for the best-fit plane,

$$Z = a + (\tan \theta_x)X + (\tan \theta_y)Y$$

If grid resolution is fine relative to the roughness of the terrain, this modification is not needed.

Gradient and aspect are only two of many possible local operators. With elevation grids, slope can be computed for any given direction, as, for example, along the trend of a proposed new road, and surface area may be estimated to reflect better the exposure of land to weathering and erosional processes. A surface can be smoothed by recomputing each elevation as the weighted average of neighboring elevations. Volumes can be measured between an existing and a planned land surface so that civil engineers can estimate the cost of cutting and filling for a proposed limited-access highway. Unwarranted irregularities captured during the optical scanning of existing maps can be identified and removed. Thick lines can be thinned, and erratic linear features smoothed.

Some local operators require more than a simple computational "pass" over each cell; in some cases the entire map must be examined sequentially, row by row and column by column, not once, but several times until an acceptable result is obtained. Multiple passes are especially common for operations logically involving a series of intermediate steps. For example, a map showing areas within 50 m of a highway can be generated by a "front" advancing stage by stage outward from an initial raster-format highway network. This type of local operator is particularly useful in environmental protection studies to define areas likely to be affected by proposed highways and utility lines. IMGRID, the Information Manipulation system for GRId cell Data structures developed by landscape architect David Sinton, can also assign suitability weights based on proximity to another cell of a particular type, such as a school, hospital, or nuclear-power plant.[5] Geographic data systems with these search capabilities can be of considerable use as analytical tools, particularly if interactive operation is provided.

A gridded map is a generalized map, with the degree of generalization proportional to the land area represented by a grid cell. If the terrain is rough, relatively large grid cells will encompass many different elevations. Slopes and aspects measured under such conditions might well be comparatively meaningless spatial averages hiding a more complex surface geometry. Moreover, with coarse grids these measurements might be altered radically by shifting the origin of the grid or varying its orientation. Useful results require cells somewhat smaller than the smallest land unit of interest; a cell size adequate for a study of field crops might not accommodate, for example, an inventory of building lots. Designing an effective grid-based system demands a thorough preliminary analysis

---

[5]David F. Sinton, *The User's Guide to I.M.G.R.I.D., An Information Manipulation System for Grid Cell Structures* (Cambridge, Mass.: Harvard University, Department of Landscape Architecture, 1977).

of likely uses and the spatial variations of the relevant phenomena. In some cases, the appropriate amount of generalization will be determined by the scale and type of graphic display, as discussed in the next section.

GRAPHIC DISPLAY

Raster data can be displayed in map form in a variety of ways by line printers, raster-mode CRT and COM units, and matrix printers. Conversion to vectors after storage, retrieval, and analysis in raster mode permits cartographic display with a pen plotter or vector CRT. Chapter 5 describes vector-mode graytone area symbols, which can represent magnitudes for grid cells, and the fitting of smoothed contours to gridded data. Additional facets of preparing raster data for mapping are presented next.

Hill Shading

Pictorial symbols can improve maps intended for general audiences unfamiliar with symbols such as contours: persons untrained in reading contour elevations and inferring slope from the spacing of contour lines can readily comprehend a land surface portrayed on a pictorial map. Oblique views of a landscape, as might be seen from the window of a low-flying aircraft or a scenic overlook, usually require the fine graphic detail provided by vector-mode display. For many purposes, oblique views, which frequently require that some lower areas located beyond ridges and peaks be hidden, are impractical because no single viewing direction is fully adequate. Completely vertical views, which obviate the hidden surface problems of oblique views, must rely upon (1) contour lines illustrating surface form, (2) usually crude and almost always highly generalized "artistic" symbols, including the hachures and "wooly worm" mountain ranges used through most of the nineteenth century and the pictorial landform symbols of skilled cartographer and physiographer Erwin Raisz, and (3) graytone symbols, which can portray either elevation classes, as with SYMAP, or the pattern of reflected light that might be seen by a viewer directly overhead in a high-flying aircraft or satellite.

Shaded relief maps are not new, and attempts to portray terrain planimetrically with tonal differences have interested many cartographers. By convention the map is shaded as if illuminated by a light source to the west or northwest of a map viewer facing north. Slopes facing northwest thus would receive and reflect more light than slopes facing southeast because an angle of incidence closer to 90 degrees concentrates more light upon a given part of the map (Fig. 4-8, left). Slope angle is important as well, because gradient influences not only the intensity of illumination on the inclined three-dimensional surface, but also the intensity of the light sensed by the viewer for each grid cell of a flat map of the region.

The darkness of the shading for a cell in an elevation grid can be inversely proportional to the amount of light L striking the inclined plane over a strip of surface with unit width oriented in the direction of the light source (Fig. 4-8, right). The corresponding length

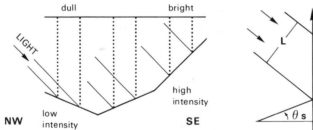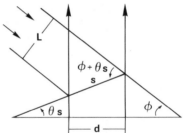

Figure 4-8 Amount of light L for unit length d on a horizontal map plane is related to the slope angle θ of the terrain and the angle of elevation φ of the light source. Surface sloping away will receive and reflect less light than a surface facing the light source.

s of the surface above horizontal strip length d is related to the inclination angle θ_s of the surface in a direction facing the light source, as given by

$$s = \frac{d}{\cos \theta_s}$$

The amount of light illuminating this expanse of the inclined plane is computed as

$$L = s \sin (\phi + \theta_s) \cdot$$

where ϕ is the elevation angle of the light source above the horizon. The directional slope θ_s is a simple function of the gradient $\nabla(X, Y)$ and the smallest angle α between directional vectors in the horizontal plane pointing toward the light source and in the direction of steepest descent; inclination angle θ_s can be found as the trigonometric inverse of its tangent, where

$$\tan \theta_s = \nabla(X, Y) \cos \alpha$$

These formulas can be combined with those given earlier for the gradient and used to compute graytone intensities for a shaded relief representation of a digital terrain elevation grid. Corrections can be made for the "atmospheric perspective," whereby the mapped pictorial view is made more realistic by exaggerating tonal contrast at higher elevations lying closer to the viewer.[6] The map can be displayed on a line printer, although the result is likely to be poor visually unless printed with a special set of geometrically simple characters providing a more stable gray scale that can portray subtle variations

---

[6]Kurt Brassel, "A Model for Automatic Hill-Shading," *American Cartographer*, 1, no. 1 (April 1974), 15–27.

Figure 4-9 Special line-printer symbols shown in Figure 3-1 can represent the form of the land surface when assigned to a display grid in accord with hill-shading intensities computed from raster data. (Courtesy Kurt Brassel.)

in illumination (Figs. 3-1 and 4-9). Hill shading is particularly appropriate for laser-beam plotters, which can produce dot-screen area symbols with a fine resolution. This pictorial representation of terrain can then be added in a visually recessive light gray or brown to an otherwise less effective map containing contour lines and standard cultural information.

Grid-to-vector Conversion

In many situations, the need for esthetically acceptable maps requires conversion of boundaries and other linear features from a raster format to a vector representation for plotting. Conversion from grid coordinates to plane coordinates is also inherent in analytical techniques requiring feature tracking, for example, measurement of the perimeter of a region or the lengths of roads and rivers.

A line can be followed relatively easily if it is thinned, that is, if every grid cell representing the feature adjoins exactly two other cells belonging to the feature, one in each direction. A stepwise tracking algorithm thus need view only a 3 by 3 subgrid around each new cell found for the feature. With a thinned line, of the eight cells surrounding the center of the subgrid, only two will have the parity code for the line: one would have been added previously to the list of cells representing the feature and the other cell is the new one to be included. Of course, a new point need be added to the vector list of nodes only when a change of direction occurs. If a linear feature is not thinned, a thinning operator would normally be applied to the map grid before tracking, although either side of a thick line can be tracked as a boundary between regions.

Tracking a regional boundary requires only a 2 by 2 subgrid, because the feature follows not a chain of grid cells, but a path between adjoining cells. Four such paths meet at the center of a 2 by 2 subgrid: the entering path traced thus far and three diverging candidates for its continuation. In tracking a boundary separating two specific regions, the new edge will also separate these regions. In tracking a boundary around one given region, the new edge will separate the specified region from a different region. Boundaries separating pairs of adjoining areas can be stored separately and the nodes at which three or more boundaries, and three or more areas, meet noted as well.

This procedure for following boundaries with a 2 by 2 subgrid assumes that the boundary has been smoothed to prevent the uncertainties presented by small, one-cell "peninsulas" and that all irrelevant enclaves and exclaves have been removed. An irregular border, perhaps where the boundary coincides with a stream with tight meander loops, must be tracked carefully to avoid making a full turn and retracing a previously recorded path in the opposite direction. A linear feature identified with a 2 by 2 subgrid has line segments oriented parallel to either the rows or the columns, and the resulting vector representation might be unduly jagged, requiring many nodes and occupying considerable memory. Unless the resolution of the data is coarse and these irregularities must be portrayed as they occur, for example, when 1-km^2 grid cells are mapped as identifiable squares 2.5 cm (0.985 in.) on a side, the esthetic appearance of the map can be improved and the cost of later processing reduced by transforming these lengthy strings of short, jagged vectors into a smoother, more streamlined chain of longer straight-line segments. Diagonal trends can be identified easily, as can long, straight vertical and horizontal lineations. Reasonable approximations to shorter trends with other orientations might be obtained by attempts at matching "template" subgrids containing prototype trends with various slopes. After these basic steps to reduce the number of nodes in vector chains, a curve generalization algorithm (see Chapter 8) might be invoked for further filtering and smoothing.

ANALYSIS OF SATELLITE IMAGERY

Land-cover information sensed from an orbiting satellite or high-altitude aircraft is an important and comparatively new source of raster-mode cartographic data. Continuous monitoring of the earth's surface at regular intervals provides a means for studying and controlling processes as slow as the spread of urban settlement or as rapid as the spread of vegetative blight. The area represented by a single *pixel* or *pel,* the grid cell of remotely sensed scanner data, is declining with the introduction of new technology. As more detailed features can be discerned from more numerous satellites with a wider range of sensors, the amount of raster-format information to be stored, processed, distributed, and displayed will increase further.

Three aspects of the cartographic use of remotely sensed data are particularly relevant to automated mapping: the transformation from digital multispectral imagery to a land-use and land-cover classification, the correction of geometric distortions in the data grid, and the production of color-composite maps from the digital classification tapes.

Radiometric Classification

Digital data received at ground tracking stations from a multispectral scanner (MSS) on an orbiting satellite consist of numbers representing the intensity of radiation energy recorded for several specific portions, or *bands,* of the electromagnetic spectrum. Although the satellite may be continually sending these radiometric observations to the network of tracking stations, the data are formated for distribution into *scenes* covering large, approximately rectangular regions. A sensor such as the multiband Thematic Mapper designed for Landsat-4 thus covers the scene of 185 by 185 km (115 by 115 miles) with four or more fine-grained and perfectly registered grid overlays. A somewhat coarser grid, covering the same area but with only one-sixteenth as many cells, is included for recording far-infrared (heat) radiation.

Individual bands provide useful information about specific phenomena, such as a differentiation of land from water, yet the full utilization of multispectral data for general-purpose mapping calls for an integrated analysis of the various grid overlays to produce a cartographically accurate and meaningful classification of pixels according to type of land use or land cover. Many types of land cover can be identified readily by their distinctive *spectral signatures,* patterns of different intensities of radiation recorded for different spectral bands. In late summer, for example, green vegatation, light-toned soil, and clear water might produce the signatures shown in the left part of Figure 4-10. If the radiometric responses were recorded for narrow bands centered on wavelengths λ_a and λ_b, clear distinctions can be made among green vegetation, which has a low reflectance for λ_a but a high reflectance for λ_b; light-toned soil, which reflects comparatively large amounts of energy for both λ_a and λ_b; and clear water, which has a low reflectance for λ_a and reflects upward virtually no radiation for λ_b. The actual spectral signatures, of course, are better represented on the graph by winding belts than by narrow lines, because the reflected energy recorded by the multispectral scanner varies somewhat depending upon atmospheric conditions and the physical condition of the target. Nonetheless, minor differences in moisture content and size, for example, would not yield signature belts so wide as to obviate discrimination among these three broad categories of land cover.

Because signatures vary somewhat depending upon the atmosphere, recent weather conditions, and, of course, seasonality, classifications must be based upon sampled prototypical land parcels experiencing the same environmental conditions as the region to be classified. These *training data* commonly are representative plots selected throughout the study area. For the land-cover types in the preceding example, the responses for bands λ_a and λ_b for some prototypical ground sites are shown in the two-dimensional plot on the right part of Figure 4-10. Boundary lines between the clusters of training data divide the graph into zones. Additional pixels can be assigned to the most nearly representative cluster by these boundaries.

Assignment to clusters on the basis of responses for more than two dimensions is difficult to illustrate graphically. Most classification techniques employ a variation of the statistical method *linear discriminant analysis.* With two spectral bands, for example,

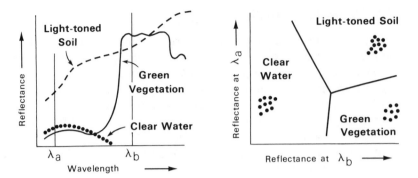

Figure 4-10 Spectral signature for three distinct land covers can be sampled at selected wavelengths λ_a and λ_b (left) and used to define clusters (right) for classifying additional pixels.

the linear boundary between two zones can be represented by the linear discriminant function

$$A\lambda_a + B\lambda_b + C = 0$$

which can be used to determine whether the land cover of a pixel is closer to cluster I than to cluster II. Three discriminant functions are required for three categories, one for each boundary. Thus, a pixel closer to cluster I than to cluster II might be assigned to cluster III if a subsequent I to III comparison indicates greater proximity to group III than to group I. New observations can be assigned mathematically rather than graphically to the most representative cluster by a computer algorithm. A complete analysis would require categories to which any significant type of land cover occurring within the scene might accurately and logically be assigned. Thus, the use of additional, carefully selected spectral bands should permit the strong discrimination among the ten or more land-cover classes needed if the map is to be generally useful.

Because some differences can be diagnosed more reliably than others, land-cover and land-use classifications often have two or more levels. For example, a major distinction at level I of one of several classifications used by the U.S. Geological Survey merely differentiates among urban, agricultural, forest, and six other general categories that might reliably be identified with satellite imagery. At level II the improved resolution possible with high-altitude aircraft is recognized by, for example, subdividing the urban category into residential; commercial and services; industrial; transportation, communications, and utilities; industrial and commercial complexes; mixed urban or built-up land; and "other urban or built-up land" categories. These classifications provide a national standard for land-planning data, and can be further refined at levels III and IV by using medium- and low-altitude imagery to represent land covers sufficiently detailed for the needs of states and localities.[7]

Geometric Adjustment

Land-cover assignments recorded on a classification tape generated by a clustering algorithm are of little value if the pixels cannot be registered to a geographic framework such as the meridians and parallels of a large-scale base map. Users of remotely sensed data need to know where as well as what, and the imagery recorded by a scanning system must also be corrected for a variety of distortions arising from minor flaws in the sensor, atmospheric refraction, the irregularity of the land surface, and aberrations in the movement of the satellite or aircraft. Although some Landsat satellites are about 900 km (520 miles) above the ground, rather than an approximately planimetric projection, the resulting view is largely an oblique perspective, with a progressively smaller scale for pixels farther from the ground track.

The basic adjustment assigns geographic or UTM grid coordinates to pixels so that the picture can be used as a map. A common approach for high-altitude multispectral scanner images taken from fixed-wing aircraft is the "rubber sheet transformation," or rectification, whereby a pair of equations is developed to relate grid coordinates for the scanner data to a map grid. For example, a somewhat crude transformation from tape pixel coordinates (x_t, y_t) to a corrected geographic grid might begin with a rotation of the image about some pivot (x_p, y_p) through angle θ to adjust for scan lines not oriented precisely north-south or east-west. The necessary pair of equations is

$$x = (x_t - x_p) \cos \theta + (y_t - y_p) \sin \theta$$

and

$$y = -(x_t - x_p) \sin \theta + (y_t - y_p) \cos \theta$$

The intermediate coordinates (x, y) are then adjusted for further systematic distortion by

$$x_g = a_1 x + b_1 y + c_1$$

and

$$y_g = a_2 x + b_2 y + c_2$$

where a_1, a_2, b_1, b_2, c_1, and c_2 are estimated according to least-squares or some other statistical method from a sample of pixels at highway intersections and other features

---

[7]James R. Anderson and others, *A Land Use and Land Cover Classification System for Use with Remote Sensor Data* (Washington, D.C.: U.S. Government Printing Office, Geological Survey Professional Paper No. 964, 1976).

more easily identified on scanner imagery. More complex polynomial equations and a
large representative sample of ground control points should permit an adjustment com-
plying with standards of horizontal accuracy for large-scale base maps.

For imagery from orbiting satellites, this empirical approach might be replaced by
the Space Oblique Mercator Projection, developed by engineer John Snyder.[8] This projection provides a continuous distortion-free mapping of satellite imagery along the ground
track, and can be used to improve the precision with which geographic coordinates are
assigned to pixels. Maps requiring compilation from several scenes can be fitted together
to cover a large region, and overlays of imagery representing different seasons or years
can be registered for a more thorough analysis that is multitemporal as well as multispectral.

A simpler adjustment problem, but one particularly troublesome with MSS data from
Landsat, is the skewing of the pixels and grid caused by the rotation of the earth beneath
the satellite. During the 25 s taken by Landsat-2 to cover the ground area on one scene,
the earth moves eastward and the ground track drifts westward one pixel width every

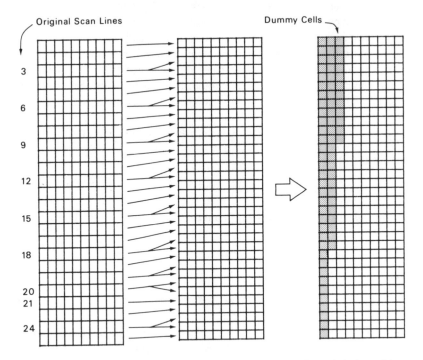

Figure 4-11 Adjustments of Landsat pixel grid for elongation of pixels and for
skew caused by earth's rotation.

[8]John P. Snyder, "The Space Oblique Mercator Projection," *Photogrammetric Engineering and Remote
Sensing,* 44, no. 5 (May 1978), 585–96.

0.124 s. A rough adjustment suitable for many applications adds dummy cells to the beginning and end of the scan-line rows to effect a convenient rectangular format for the entire grid (Fig. 4-11). Additional rows are also added, by duplicating every third and twentieth scan line, to correct for pixels 57 m (190 ft) wide by 79 m (259 ft) long. The resulting scene-corrected magnetic tape, converted from 2,340 scan-line records of 3,240 pixels each to a grid with 3,237 rows of 3,500 square cells each, is more appropriate for further analysis and cartographic processing.[9]

Color Composite Maps

Scanner data are digital, not photographic, and the often vivid color composite prints made from these data resemble photographs only because the digital data have been converted to color images at scales small enough to obscure the grain of the pixel grid. Digital reflectances can be converted to a single color negative for standard photographic reproduction or to color separation negatives for plate etching and offset printing. Because there is no inherent correspondence between the photographic dyes or printing inks of the reproduced image and the infrared "colors" sensed by a scanner, considerable experimentation by cartographers is possible with the numerous composites that might be produced.

A typical color product is the composite image produced by exposing Landsat MSS bands 4 (green), 5 (red), and 7 (the longer of two near infrared bands) in registration and with filters onto color film. The resulting prints have the conventional spectral shift of false-color imagery: targets with a dominant response in the green band appear blue, those with a dominant response in the red band appear green, and those, such as healthy vegetation, with a dominant response in the infrared band appear red (Fig. 4-12). Objects such as blue water, with minimal reflectance in these three bands, appear black.

Enlarged prints at scales of 1:250,000 and 1:500,000 provide useful views of regional land cover. Color separation negatives for lithographic printing can be obtained by enlarging photographic negatives exposed in a laser-beam plotter driven by a geometrically corrected computer tape. A map grid, appropriate labels, and such cartographic enhancements as political boundaries and major highways can be added during printing on a black-ink plate in registration with the color plates. Several scenes can be combined for a mosaic of a state or region. With suitable geodetic control, the computer preparing a tape for a laser plotter can make the geometric adjustments needed for the smooth graphic transition between adjoining scenes mapped to a common projection.

The cartographic enhancement of MSS imagery might, of course, involve a land-use and land-cover classification. In this case the mapped information is qualitative, rather than quantitative, and the spectral shift of color infrared imagery must be exchanged for easily differentiated color symbols portraying numerous land-cover types. On some maps

---

[9]A. N. Williamson, "Corrected Landsat Images Using a Small Computer," *Photogrammetric Engineering and Remote Sensing*, 43, no. 9 (September 1977), 1153–59.

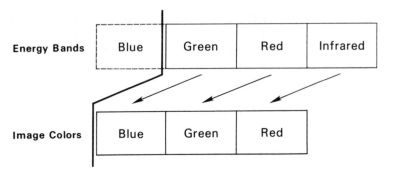

Figure 4-12 The spectral shift of color infrared imagery displays as red areas reflecting largely infrared energy and thus bumps downward the red and green bands. Blue radiation is not recorded.

the enlarged shapes of individual pixels may even be evident because land-cover information is particularly useful at scales larger than 1:250,000. A laser-beam plotter randomly positioning halftone dots on the various film separates used in offset printing can yield maps with many different tones and hues needed for a qualitative distribution with many categories.

Interactive Systems

Classification algorithms based on linear discriminant analysis and using training data for large numbers of land-cover categories are executed efficiently on an attached array processor or on powerful, large-memory parallel processors such as the ILLIAC-IV and its successors at NASA's Ames Research Center. Nonetheless, researchers familiar with their areas and phenomena can usually obtain suitable results, quickly and when needed, with an *image analyzer,* an interactive microprocessor with a color CRT display. Classifications can be developed empirically through *density slicing,* whereby digital responses in particular bands can be assigned to classes based on thresholds. To illustrate, category zones similar to those in Figure 4-10 might be formed by boundaries perpendicular to the axes at threshold reflectances determined through trial and error. Reflectance responses commonly have six- or seven-bit accuracy, allowing 64 or 128 possible levels with 63 or 127 different thresholds for each band. Experimentation with different thresholds, for example, to distinguish healthy plants from strawberries attacked by insects or fungi, might produce a satisfactory classification for a small window familiar to the investigator. A color CRT display is particularly useful for evaluating patterns produced by various thresholds, because the analyst can assign primary hues to give a strong visual definition to raster boundaries and lessen the likelihood of misinterpretation. By interacting with the computer through the CRT, the analyst can experiment with these thresholds, test them elsewhere in the region, and adjust them if necessary.

5

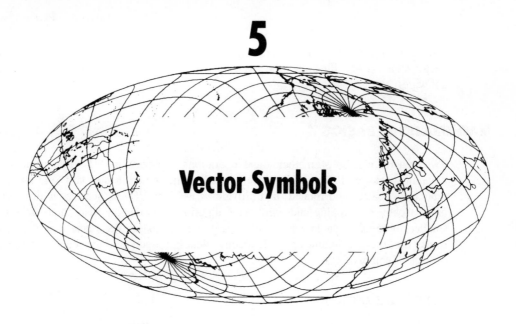

Vector Symbols

Vector symbols are related more closely than raster symbols to traditional cartographic practice. Maps drawn by hand with an ink pen can be represented by long lists of (X, Y, P) coordinates, where X and Y are plane coordinates and P is a pen code specifying line width or, if zero, the pen-up position. Most point, line, and area symbols are readily adapted to a vector format. For example, the dots that collectively show variations in density on a dot-distribution map can be specified individually by two entries in a vector list: the first entry positioning the pen and the second dropping the pen to mark the dot. Similarly, dashes can be drawn by using different coordinates for the second point and a pen code to drop the point before movement of the pen. Orderly assemblages of these dashes or dots can produce a line symbol or an area symbol.

Inexpensive pen plotters permit the automated drafting of publishable maps, and flatbed plotters equipped with a lighthead can expose directly the film separates for color maps. Vector-mode CRT displays provide rapid previewing and interactive editing, and vector-mode laser-beam plotters can produce graphic-arts quality labels and symbols for color separations from which printing plates can be etched. For many users, vector-mode devices are the most versatile available display hardware producing esthetically acceptable computer-assisted maps.

More important than the ease with which modern display hardware can plot points and line segments is the facility with which vector symbols can be manipulated by the computer. Moreover, vector symbols can portray raster data and the results of raster-mode analyses. Mapping software is available for almost every useful map projection and cartographic symbol, and experiments indicate that vector data can be smoothed and filtered rapidly and precisely. This chapter explores the mathematical principles underlying vector symbols and illustrates the uses and limitations of these principles in mapping software.

MATHEMATICAL BASICS

Two general classes of mathematical operations are basic to a flexible vector-mode mapping package. Point, or instance, transformations permit standard symbols to be replicated, scaled, and rotated, and provide as well for viewing surfaces with a variety of perspectives. Clipping and windowing algorithms permit the map author to display smaller, more manageable portions of large geographic base files. Because of their widespread use, these techniques provide a convenient introduction to the scope and potential of vector symbols.

Point Transformations

Prototypes of vector point symbols can be stored in a master list as N coordinate pairs (x_i, y_i) referenced to a local origin centered within the symbol. Each time the symbol is to be plotted, these N points can be copied from the master list into the longer list containing all points for the map. The plane coordinates added to this map list are not exact duplicates of those in the master list for the symbol but a transformed set of coordinates (x_i', y_i') incremented by the coordinates (u, v) of the point on the map about which the symbol is to be centered. This transformation is called a *translation* and is specified by the equations

$$x_i' = x_i + u$$

and

$$y_i' = y_i + v$$

Quantitative symbols must be enlarged or reduced so that symbol size represents magnitude of the distribution at mapped locations. Examples include graduated circles, with a vector list of many short straight-line chords approximating the smooth circumference of a perfect circle; shorter lists, such as graduated squares and vertical bars; and relatively complicated lists requiring several pen-up and pen-down commands, as for a pictorial symbol of a sheath of wheat portraying the size or value of a harvest. The scaling needed for proportional point symbols is added to translation in the transformations

$$x_i' = sx_i + u$$

and

$$y_i' = sy_i + v$$

where scaling coefficient s will enlarge the symbol if greater than 1 and shrink the symbol if less than 1.

Graduated point symbols are usually scaled so that the apparent area of the symbol is proportional to a magnitude, such as the population of a city or the value added by all manufacturing plants within a county. Scaling can be relative to a standard symbol, part of the map legend. For example, if a standard circle with radius r_s is to represent magnitude z_s, scaling factor f, computed as

$$f = \frac{r_s}{z_s^{1/2}}$$

can be used to find radii r_k for other point-symbol magnitudes according to

$$r_k = f z_k^{1/2}$$

Because map viewers tend to underestimate the sizes of larger circles, and hence the larger magnitudes in geographic distributions, an exponent larger than $1/2$ is often used to compensate for this phenomenon. Circles portraying the larger magnitudes are enlarged further so that the resulting underestimated magnitudes more nearly approximate the true values. Attempts to estimate appropriate exponents for this type of *apparent-value rescaling* frequently employ subject-testing methods from perceptual psychology and psychophysics.[1]

Point symbols can be rotated about a coordinate origin by the transformations

$$x_i' = x_i \cos \theta - y_i \sin \theta$$

and

$$y_i' = x_i \sin \theta + y_i \cos \theta$$

where θ is the counterclockwise angle that brings the original points into the desired orientation. Rotation is particularly common with symbols such as arrows depicting locally important wind directions, directional aspects of flows of migrants and manufactured goods, and slope directions.

If a prototype point symbol with a local coordinate origin is to be scaled and translated as well as rotated, scaling and rotation may occur in any order, but both operations must precede the translation. Otherwise, the point symbol might inadvertently be shifted and stretched toward or away from the map origin pivot or tossed to an unwanted position elsewhere within, or possibly outside, the map window.

If the coordinates of a symbol are not based upon axes with an origin at the desired pivot, a translation of the symbol center to the origin must precede the rotation, and a

---

[1]See, for example, Judy M. Olson, "Experience and the Improvement of Cartographic Communication," *Cartographic Journal*, 12, no. 2 (December 1975), 94–108.

translation back to the original position must follow. The appropriate formulas for local pivot point (x_p, y_p) are

$$x_i' = [(x_i - x_p) \cos \theta - (y_i - y_p) \sin \theta] + x_p$$

and

$$y_i' = [(x_i - x_p) \sin \theta + (y_i - y_p) \cos \theta] + y_p$$

Matrix Operators

Rotation, scaling, and translation can be represented by simple matrixes, 2 by 2 arrays of numbers that can be multiplied by the original coordinates to yield the desired transformation. The appropriate general form of matrix multiplication is represented symbolically by

$$C = A B$$

where A and C are one-row-by-two-column matrixes called *row vectors* and B is a 2 by 2 matrix operator. The two elements of product matrix C are computed as

$$(c_1 \quad c_2) = (a_1 b_{11} + a_2 b_{21} \quad a_1 b_{12} + a_2 b_{22}) = (a_1 \quad a_2) \begin{pmatrix} b_{11} & b_{12} \\ b_{21} & b_{22} \end{pmatrix}$$

The first element of C is the sum of the products of the elements of A multiplied by the corresponding elements in the first column of B. The second element c_2 is a similar sum of products for the second column of B. In FORTRAN programs, these finer details of matrix arithmetic can be ignored by a call in the form CALL MXMULT (A, B, C, 2, 1) to a packaged subroutine. An extended version of BASIC requires no mention of the numbers of rows and columns of the product matrix in the corresponding statement MAT C = A * B, and the matrix-oriented interactive language APL permits a similar computational shorthand in the command C←A + . × B.

   Matrix notation is particularly useful in written descriptions of algorithms if, instead of Roman and Greek letters, transformation operators are given more easily interpreted labels such as Rot, Scl, and Trl for rotation, scaling, and translation. Further convenience requires converting the arithmetic required for translation from addition to multiplication, as used for both rotation and scaling. This simple conversion adds an additional element to each row and column vector and to each row and column of the transformation matrixes. With vectors representing coordinate pairs the additional element is a 1, which yields the two-dimensional homogeneous coordinates shown for original point vector $X(i)$ and transformed point vector $X(i)'$ as

$$X(i) = (x_i \quad y_i \quad 1)$$

and

$$X(i)' = (x_i' \quad y_i' \quad 1)$$

A translation to point p is thus given by the matrix multiplication

$$X(i)' = X(i) \, \text{Trl} \, (x_p, y_p)$$

where the elements of the translation operator are arrayed as

$$\text{Trl} \, (x_p, y_p) = \begin{pmatrix} 1 & 0 & 0 \\ 0 & 1 & 0 \\ x_p & y_p & 1 \end{pmatrix}$$

The corresponding arrangements for the elements of operator Rot (θ) performing a counterclockwise rotation through angle θ is

$$\text{Rot} \, (\theta) = \begin{pmatrix} \cos\theta & \sin\theta & 0 \\ -\sin\theta & \cos\theta & 0 \\ 0 & 0 & 1 \end{pmatrix}$$

The scaling operator Scl (s) incorporates scaling factor s in the first two diagonal elements, as shown by

$$\text{Scl} \, (s) = \begin{pmatrix} s & 0 & 0 \\ 0 & s & 0 \\ 0 & 0 & 1 \end{pmatrix}$$

These 3 by 3 matrixes can be multiplied by each other and the translation operator to yield a 3 by 3 product matrix incorporating two or more types of transformation. Moreover, the row vector of original coordinates can be replaced by an N by 3 matrix, with N rows of coordinates for N separate points, so that the result of matrix multiplication is an N by 3 matrix containing the transformed coordinates in its N rows. The third coordinate in each row cannot be used as a pen code and will retain the value 1.

Consider the arrow shown in Figure 5-1 and represented by the 5 by 3 coordinate vector X and the pen-code column vector P, given as

$$X = \begin{pmatrix} 1 & 2 & 1 \\ 2 & 2 & 1 \\ 2 & 1 & 1 \\ 2 & 2 & 1 \\ 0 & 0 & 1 \end{pmatrix} \qquad P = \begin{pmatrix} 0 \\ 1 \\ 1 \\ 0 \\ 1 \end{pmatrix}$$

The arrowhead is specified by the first three points (rows in X), and the pen must be lifted for the fourth point, the juncture of the head and shaft. The initial direction of the arrow is toward the upper right. Subsequent positions shown in Figure 5-1 are the cumulative results of the following:

1. A clockwise rotation through 90 degrees, specified as a negative counterclockwise angle in Rot $(-90°)$.

2. Shrinkage to half the original size by operator Scl (0.5).

3. Translation to $(3, 2)$, the new coordinates of the tail of the arrow given in Trl $(3, 2)$.

4. A counterclockwise rotation through 90 degrees, specified by Rot $(90°)$.

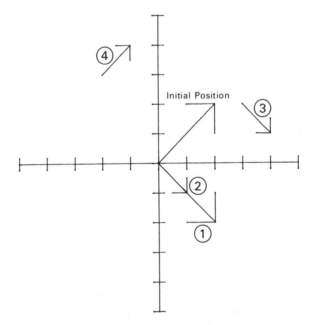

Figure 5-1 Sequence of arrows produced by (1) a quarter clockwise rotation, (2) shrinkage to half scale, (3) translation to point (3, 2), and (4) a quarter counterclockwise rotation.

These successive transformations could be included in a single matrix equation, with later operators added, in order, at the right.

The first operator is toward the left, next to the matrix of original coordinates, and the order of the operators is crucial. For example, the equation

$$X' = X \text{ Rot } (-90°) \text{ Scl } (0.5) \text{ Trl } (3, 2) \text{ Rot } (90°)$$

produces the sequence of arrows shown in Figure 5-1, but the chain of operators in

$$X' = X \text{ Rot } (-90°) \text{ Rot } (90°) \text{ Scl } (0.5) \text{ Trl } (3, 2)$$

yields an arrow pointing directly upward from a tail anchored at (3, 2).

In this last equation the first two operators to the right of the matrix of initial coordinates have no combined effect: a quarter-counterclockwise rotation undoes the effect of the immediately preceding quarter-clockwise rotation. This example illustrates an inverse transformation. The inverse $\text{Rot}^{-1}(\theta)$ of a rotation Rot (θ) is a rotation Rot $(-\theta)$ of the same magnitude but in the opposite direction. Similarly, the inverse translation $\text{Trl}^{-1}(x, y)$ is simply translation Trl $(-x, -y)$ with the signs of the coordinates reversed. Because division is the inverse of multiplication, the inverse scaling transformation $\text{Scl}^{-1}(s)$ is merely scaling operator Scl $(1/s)$ using the inverse $1/s$ of the original scale factor s. These transformations and their inverses are particularly useful with an interactive graphics system, on which a map might be composed by an operator or map author scaling, rotating, and translating point symbols to positions specified with a light pen. As another example, symbols plotted on maps covering large portions of the world might also be rotated individually by mapping software to conform to local orientations of meridians and parallels. At substantially larger scales, for example, on engineering plans for a new subdivision or urban redevelopment project, symbols representing utility connections to structures can be added with an appropriate orientation to building symbols aligned with curving streets.

Windowing and Clipping

Cartographic data bases commonly contain a variety of geographic features for large regions, yet individual maps often address specific themes for limited areas. Considerable computational effort can be required to separate line segments lying wholly within a given window from those falling outside the area of interest. Segments crossing the map border must be clipped by replacing the outside node with the intersection of the segment and the window. The appropriate side of the map frame must be found, and in the case of a chain of line segments enclosing an area, a part of the map's rectangular boundary must be inserted in the chain to assure closure. When irregular polygons intersect the border, a single polygon might yield several fragments, each a separate closed chain of segments but linked nonetheless to the feature characteristics recorded for the original area.

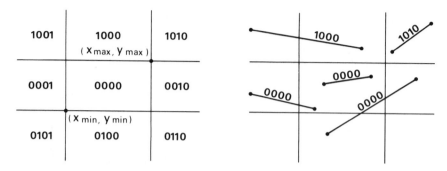

Figure 5-2 Four-condition bit strings of the nine regions (left) in which end points might lie can be combined with bit-by-bit multiplication into a composite string of bits representing the entire segment. Segments with any non-zero bits lie entirely outside the window.

Efficient identification of line segments falling entirely outside the window simplifies the windowing process.[2] Each end point of a line segment is evaluated for four conditions:

1. Point above upper border.
2. Point below lower border.
3. Point to right of right-hand border.
4. Point to left of left-hand border.

Each condition is represented by a binary digit set to 1 if the condition holds and to 0 otherwise. The resulting string of four bits represents one of nine regions produced by extending indefinitely the borders of the rectangular window (Fig. 5-2, left). Each condition is tested by comparing one coordinate of the end point with the corresponding coordinate of an appropriate corner of the frame. For instance, an end point lying above the window would be identified as satisfying the first condition because its y coordinate exceeds the y coordinate y_{max} of the upper-right corner.

A new bit string is then developed for the entire segment by setting each condition bit to 1 if and only if the corresponding bits for the two end points are both 1. If any bit of this composite bit string is a 1, the line segment lies entirely outside the window and can be discarded. Moreover, if both original bit strings for the end points are entirely 0's, the segment can be plotted without clipping.

A line segment identified as neither entirely excluded nor completely included must be clipped either once, if an interior end point is recognized by a string of zero condition bits, or twice, when the window lies between external end points (Fig. 5-2, right).

---

[2]William M. Newman and Robert F. Sproull, *Principles of Interactive Computer Graphics* (New York: McGraw-Hill Book Company, 1973), 123–26.

Although other, more computationally efficient algorithms have been proposed, a direct solution to the clipping problem finds all intersections between extensions of the line segment and the borders of the window, and then determines which intersections lie between the end points of the segment (Fig. 5-3, left). If more than the expected one or two intersections remain, an additional test is applied to detect and save only points along the edge of the window. These tests merely compare one of the intersection coordinates with an appropriate range of the same coordinate along the line segment or border (Fig. 5-3, right). Unless a line segment is vertical, an intersection with a border lineation would be recognized as lying within the segment if the x coordinate of the intersection were neither less than the smaller nor greater than the larger of the x coordinates of the end points.

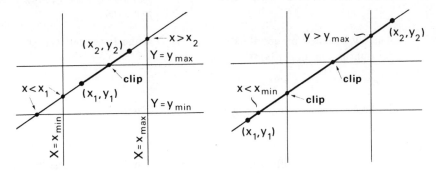

Figure 5-3 Intersections for clipping remain after discarding intersections beyond the line segment (left) and outside the window (right).

Intersection coordinates can be computed by representing the line segments and borders with linear equations and obtaining the simultaneous solution for the point satisfying each set of equations. The basic form of a linear equation is

$$Y = a + bX$$

where a is the y intercept and b is the slope (Fig. 5-4). For a line segment with end points (x_1, y_1) and (x_2, y_2) the slope can be computed as

$$b = \frac{y_1 - y_2}{x_1 - x_2}$$

Once the slope is determined, the y intercept is computed by substituting the slope into the linear equation and rearranging terms for the calculation of a according to

$$a = Y - bX$$

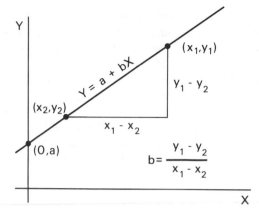

Figure 5-4 Slope b and Y intercept a, parameters in the equation for a straight line, can be computed from the coordinates of two points on the line.

A window bounded by coordinates (x_{min}, y_{min}) at the lower left and (x_{max}, y_{max}) at the upper right has an upper border aligned along

$$Y = y_{max}$$

and a lower border aligned along

$$Y = y_{min}$$

The intersection of the line segment, or its extension, with this lower border can be found by combining the linear equations for segment and border to solve for the x coordinate as

$$x_{int} = \frac{y_{min} - a}{b}$$

This intersection, of course, has the same y coordinate, y_{min}, as the lower border.

Similarly, the left- and right-hand borders of the window, represented by

$$X = x_{min}$$

and

$$X = x_{max}$$

provide the x coordinates of their intersections with the map feature. Direct substitution of the x coordinate yields the corresponding y coordinate for the intersection, as given by

$$y_{int} = a + bx_{int}$$

Most map formats, including the screens of CRT display units, are rectangular, and thus can be accommodated by these relatively simple formulas, upon which the built-in windowing functions of interactive graphics systems are based. If the borders of a rectangular window are not parallel to the coordinate axes for the data base, a rotation of feature coordinates to axes parallel to the map window will simplify clipping. With nonrectangular windows, such as circles and regions bounded by curved meridians and parallels on small-scale maps, clipping is possible nonetheless, although the calculations are more complex.

SYMBOL GENERATION

In addition to clipping and simple geometric transformations, analytic geometry and linear algebra are useful in the derivation of numerous symbols that might be used to represent geographic features represented as vector data. An appreciation of a few basic examples of these geometrically derived symbols provides added insight to the potential of computer-assisted cartography.

The Railroad Symbol

A linear symbol with short, evenly spaced cross-ticks is used on large- and small-scale maps to represent railways. The computations for the automated plotting of this symbol demonstrate the measurement of distance along a feature, as well as the construction of other symbol elements, in this case the cross-ticks, based upon the slopes of the line segments defining the feature. A wide variety of traditional boundary symbols composed of periodically repeated dots or dashes of different lengths is also plotted by "pacing off" distance along a linear feature. Slope must be considered in generating perpendicular feature elements, such as the railway cross-ticks and the ticks on the downslope side of depression contours, and parallel features, such as parallel lines representing residential streets and the double-dash lines symbolizing unimproved dirt roads.

Distance along a linear feature is computed by accumulating the lengths of straight-line segments, in sequence from a designated starting point. Direct distance d_{12} between points (x_1, y_1) and (x_2, y_2) is computed as

$$d_{12} = [(x_1 - x_2)^2 + (y_1 - y_2)^2]^{1/2}$$

Cross-ticks might occur at one or more intermediate points (x_p, y_p) located at distances d_{1p} from the first node of the segment, the node at which the segment is joined to the previous segment in the chain. These coordinates are computed as

$$x_p = x_1 + (x_2 - x_1)\frac{d_{1p}}{d_{12}}$$

and

$$y_p = y_1 + (y_2 - y_1)\frac{d_{1p}}{d_{12}}$$

After a new cross-tick is plotted for the symbol, the distance d_{1p} from the first node of the segment is incremented by the unit spacing of the ticks to yield a new distance d_{1p}'.

If d'_{1p} exceeds d_{12}, no additional cross-ticks occur along the line segment. If the next line segment is sufficiently long, the next cross-tick will occur at distance d'_{1p} from the first node of the new segment, computed as

$$d'_{1p} = d_{inc} - (d_{12} - d_{1p})$$

the tick interval d_{inc} minus the portion of the length d_{12} remaining after the last tickmark was plotted at distance d_{1p} from the beginning of the previous segment. If d'_{1p} is less than the length of the new segment, the cross-tick will occur on the next, or possibly a subsequent, segment along the rail line.

The slope b_c of a cross-tick is the negative inverse $(-1/b_s)$ of the slope b_s of the feature segment. Coordinates (x_c, y_c) of one end of a cross-tick of length $2t$ can be found for distance t from the intersection (x_p, y_p) by converting the slope b_c to the orientation angle ϕ, counterclockwise from a line through the intersection and parallel to the X axis (Fig. 5-5). Angle ϕ is the inverse tangent of the slope of the tick, computed as

$$\phi = \tan^{-1} b_c$$

and the sine and cosine of the angle yield the coordinates of the diametrically opposite end points, given by

$$x_c = x_p \pm t \cos \phi$$

and

$$y_c = y_p \pm t \sin \phi$$

Because the zero slope of a horizontal line segment would produce an infinitely large slope b_c for the cross-tick, when $b_s = 0$ the end points can be determined quite simply as $(x_p, y_p + t)$ and $(x_p, y_p - t)$.

These procedures can be extended to obtain two parallel lines with slope b_s offset an equal distance t from a center line along the feature. New nodes would need to be computed for this street symbol, because a line segment on the inside of a bend is shorter and the corresponding segment on the outside is longer than the center line segment along the feature. For lines to be scribed by a plotter removing an opaque emulsion from an otherwise transparent film with a narrow blade, the slope of a line segment can provide the plothead with the proper orientation for the cutting tool.

The Circle

Circles are used on maps as graduated point symbols, as the circumferences of pie charts showing both magnitude and proportions for various categories, as the uniformly spaced, repeated elements of area symbols for orchards and certain types of surficial deposits, and as parts of more complex point symbols representing weather conditions. A many-sided polygon can resemble a circle if the nodes are equidistant from the center. The

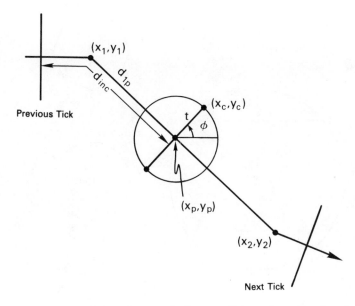

Figure 5-5 Geometry of perpendicular cross-ticks on the railroad symbol.

greater the number of sides, or chords, the smoother the circumference. Yet the number of polygon edges might be so great that considerable time is required for plotting, and with liquid-ink or felt-tip pens excessive amounts of ink might diffuse into the fibers of the paper to produce a thick, fuzzy circumference.

An appropriately rounded circle can be produced by specifying only as many chords as needed to maintain a minimum deviation d_{max} between the edges of a regular polygon and the circumference of a true circle (Fig. 5-6). In a circle of radius r, a sector angle 2θ subtends a chord no closer than $r \cos \theta$ to the center. Conversely, this chord is no farther than $r (1 - \cos \theta)$ from the circumference. The resulting relationship among maximum distance, half-sector angle, and radius, given as

$$d_{max} = r (1 - \cos \theta)$$

can be solved for half-sector angle θ as

$$\theta = \cos^{-1} \left(1 - \frac{d_{max}}{r} \right)$$

The corresponding number of chords thus can be computed as

$$n_{chords} = \text{iceil} \left(\frac{\pi}{\theta} \right)$$

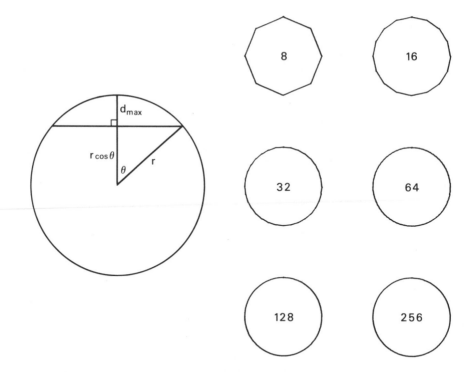

Figure 5-6 Geometry of a circle approximated by a regular polygon (left) and examples of circles approximated by different numbers of chords (right).

where angle θ is in radians and the ceiling function iceil rounds a decimal number to the next highest integer. A large circle requires more chords than a small circle to maintain the same maximum deviation between chord and true circumference.

A circle of radius r is plotted by successive rotations of the point $(r, 0)$ about the origin in small angular increments 2θ, with translation to the intended center point (x_p, y_p) before plotting. A maximum deviation of 0.025 mm (0.001 in.) generally yields regular polygons resembling smooth circles, although smaller tolerances might be needed for relatively thin circumferences. Two additional circuits of complete rotations about the center, one with a slightly larger radius and the other with a slightly smaller radius, will yield a noticeably thicker circumference, also useful if the circles are to dominate visually other symbols on the map.

Shaded Circles

Greater visual contrast between the circle and its background should focus the map viewer's attention more directly upon the distribution portrayed by the circles. Shading of the interior of a circle will generally produce a stronger figure-ground relationship than a thick circumference. Shading of the interior also permits the circle to portray ratios and percentages for point locations such as cities on a small-scale map. A set of uniformly spaced shading lines can be produced by plotting parallel chords outward from the center in opposite directions. If the first shading line is a diameter of the circle, successive chords would be plotted at distances kd_{sep}, integer multiples k of the unit line separation d_{sep}. For the two parallel chords located the same distance above and below the diameter parallel to the X axis of a circle centered at (x_c, y_c), the y coordinates of the end points are $y_c + kd_{sep}$ and $y_c - kd_{sep}$ (Fig. 5-7). Because the end points are offset equally to the left and right of the center, the x coordinates are $x_c + [r^2 - (kd_{sep})^2]^{1/2}$ and $x_c - [r^2 - (kd_{sep})^2]^{1/2}$. The offset $[r^2 - (kd_{sep})^2]^{1/2}$ is derived using the Pythagorean theorem, whereby the square of the hypotenuse of a right triangle is the sum of the squares of the other two sides.

End points can also be derived for chords parallel to the Y axis or for two perpendicular sets of chords inclined at angles of 45 and -45 degrees to the horizontal. The pattern and texture of this crossed-line shading pattern are less harsh visually than if only a single set of shading lines were used, and are less likely to distract the map viewer from the graytone of the symbol (Fig. 5-8). The proportion p of the area inked is related to plotted line width w and the separation d_{sep} according to

$$p = \frac{2d_{sep}(w/2) + 2[d_{sep} - 2(w/2)](w/2)}{d_{sep}^2}$$

which reduces to

$$p = \frac{2wd_{sep} - w^2}{d_{sep}^2}$$

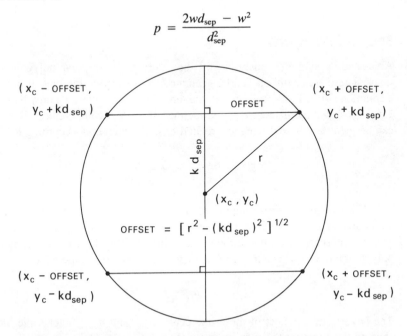

Figure 5-7 Geometry of chords used for shading the interior of a circle with parallel lines of separation d_{sep}.

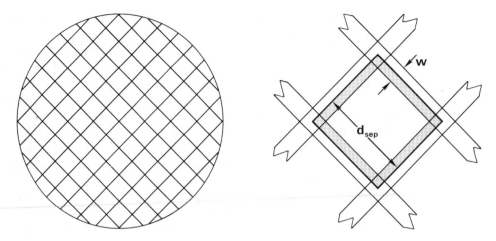

Figure 5-8 Crossed-line shading for circle interior (left) and elementary cell with parameters d_{sep} and w (right).

The separation d_i needed to represent proportion p_i can be computed as

$$d_i = \frac{w + w\,(1 - p_i)^{1/2}}{p_i}$$

Shaded Polygons

Area-shading symbols must also fill the interiors of irregular polygons on choropleth maps. These area symbols can be generated as one or two families of parallel lines with a common slope. Individual lines in the family thus are identified only by their intercepts. Lines separated by distance d_{sep} and inclined at angle θ counterclockwise from the X axis can be identified by the integer k related to the y intercepts according to

$$a_k = k\,\frac{d_{sep}}{\cos\theta}$$

The y intercept is also important in finding the intersection between the sides of the area polygon and the members of the family of shading lines. Each node of the polygon can be represented by a line through the point with the slope b_s of the shading lines. For polygon node (x_j, y_j) the intercept a_j of this line can be computed as

$$a_j = y_j - b_s x_j$$

The largest and smallest of these intercepts, a_{max} and a_{min}, bracket the intercepts of all shading lines that intersect the polygon (Fig. 5-9, left). Only shading lines having intercepts within this range need be considered.

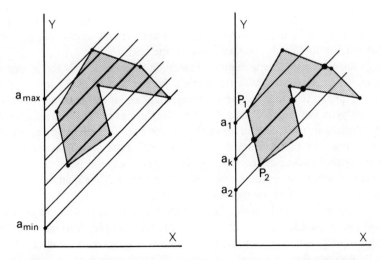

Figure 5-9 *Y* intercepts of lines parallel to shading lines and extending through the polygon nodes bracket all shading lines intersecting the polygon (left), and for each polygon side, bracket all shading lines intersecting the particular edge (right).

The intercept is also useful for determining which polygon sides intersect a particular shading line. A shading line with intercept a_k will intersect only polygon edges with one intercept greater than or equal to a_k and the other intercept less than or equal to a_k (Fig. 5-9, right). Side i of the polygon, with slope b_i and intercept a_i, will intersect the shading line with slope b_s and intercept a_k at

$$x = \frac{a_k - a_i}{b_i - b_s}$$

The linear formula for the shading lines yields the corresponding y coordinate, computed as

$$y = a_k + b_s x$$

The time required to compute shading lines can be conserved by storing, with the co-ordinates for each polygon node, the y intercepts of lines through the node having the slopes of all shadings likely to be used.

Irregularly shaped polygons often intersect the same shading line in several places, yielding two or more shading-line segments represented by four or more intersections. The correct pen commands can be generated by sorting the intersection points according to their x coordinates: the pen will be up when plotting to the point with the lowest x coordinate, down when plotting to the second point, up when plotting to the point with the third lowest x coordinate, and so forth. This approach also accommodates enclaves, which will be encountered first on their left sides, as odd-numbered intersections in a list sorted by x coordinate. The pen lifted at the second, or first even-numbered, intersection in this list will be dropped at the third, or second odd-numbered, intersection to resume the shading line on the opposite side of the enclave.[3]

In some instances a shading line might intersect a polygon at one of the nodes. A complicated set of decision rules might be established either to assign this intersection to one of the two intersecting polygon edges or, if the shading line does not penetrate the polygon farther, to ignore it. A far simpler but quite adequate solution merely shifts the shading line in question upward or to the right, depending upon the slope of the line, by ever so slight an amount that a direct "hit" at the node is avoided without noticeably altering the spacing of the shading lines. There is little point to more complicated and time-consuming procedures that, no matter how elegant mathematically, would add no visually perceptible improvement to the map.

Crossed-line and parallel-line symbols are but two of many possible area symbols that can be produced. Qualitative maps of, for example, land uses, geologic formations, climatic regions, and crop types commonly require area symbols differing largely in pattern rather than graytone. Of course, if color reproduction is to be used, all regions of the same type might, for example, be converted to a raster format and filled in on a film negative by a laser-beam plotter to provide a mask for use with tint screens and process colors. Yet, for a less elaborate map, patterned vector-mode area symbols consisting of two-dimensionally ordered squares, dashes, crosses, or more elaborate symbol elements are convenient for portraying differences in character rather than quantity (Fig. 5-10). These symbol elements can be centered and spaced evenly along unplotted "shading lines" generated solely for this positioning. Distances between prospective pattern elements to the nearest polygon boundary can be checked to avoid overlap with other area symbols. Symbol elements of various types, orientations, and sizes can be alternated to provide a greater variety of qualitative area symbols.

Contour Threading

Traditionally, contours and other isolines are drawn as vector symbols rather than portrayed implicitly as gaps between shaded areas on raster maps, as with SYMAP. In computer-assisted cartography, contour lines can be plotted as vector symbols by *threading* a chain of straight-line segments through a two-dimensional lattice of grid intersections at which surface values have been estimated by interpolation, as described in Chapter 3. A contour that enters a square cell at one side of this lattice should exit the cell through some other side. A contour representing value z_c crosses a cell side with a value at one end greater than z_c and a value at the other end less than z_c (Fig. 5-11, left). The contour can then be threaded into the next cell and so on until it either encounters the edge of the map or closes upon itself. Although rare, a contour might intersect the square cell

---

[3]For a detailed explanation of the polygon shading algorithm, see Mark S. Monmonier and D. Michael Kirchoff, "Choropleth Plotter Mapping for a Small Minicomputer," *Proceedings of the American Congress on Surveying and Mapping*, 37th Annual Meeting, 1977, 318–38.

Figure 5-10 Geometric area-shading patterns (right) plotted by a flatbed plotter, with 400 percent enlargement (left) to illustrate detail. (Courtesy U.S. Geological Survey.)

at a corner, where, of course, the estimated surface elevation would exactly equal z_c. Because uncertainty arises when two different contours representing the same elevation fall within the same square cell, lattices of triangles rather than squares are sometimes used.[4]

Simple linear interpolation can be used to locate the point at which the contour passes between the corners of a cell (Fig. 5-11, right). If the distance from corner A to corner B is s, and if the corner elevations are z_a and z_b, the contour will cross the cell side at distance d_{ac} from A in the direction of B, where

$$d_{ac} = \frac{z_a - z_c}{z_a - z_b} s$$

When all contour-grid intersections have been identified, the chain of straight-line segments representing the contour can be plotted.

Some contours at a given elevation can be initiated by inspecting the edge of the mapped area, but the certain detection of contours lying entirely within the area requires the systematic inspection of interior cell sides. Computational efficiency can be improved by recognizing that contour lines representing a continuous surface have a systematic, nested structure. For example, if the contour interval is 20 m, a 120-m contour must lie between adjacent 100- and 140-m contours. If further detail might be added later, in the form of supplementary contours at a smaller contour interval, it is highly useful to maintain

[4]J. Ross Mackay, "The Alternative Choice in Isopleth Interpolation," *Professional Geographer,* 5, no. 4 (July 1953), 2–4.

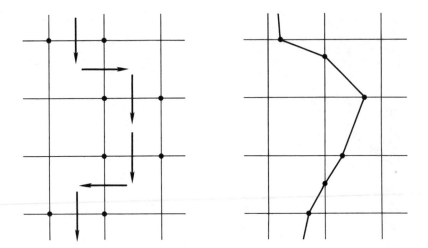

Figure 5-11 A contour is threaded through a grid by first identifying cell sides with end-point elevations bracketing the contour elevation (left) and then estimating the crossing points through linear interpolation (right).

indicators of the nesting of contour lines as a part of the data file.[5] Moreover, identification of peaks and pits in the surface is needed to assure the complete representation of knolls, depressions, and hilltops.

A cartographic draftsperson using traditional techniques would never represent a contour as a series of straight-line segments with sometimes abrupt changes in direction. In computer-assisted cartography, smoothed contours, with less jerky paths, could be obtained by using an interpolation grid so fine that the resulting chains of very short line segments would resemble curved lines. This approach would waste computer time unnecessarily, for a simpler smoothing procedure can obtain a similar result with much less computational effort.

Cubic spline interpolation is an attempt to simulate with a computer the draftsperson's use of a spline, a thin plastic strip that serves, in a sense, as a bendable "straight edge" for guiding a drawing pen along a gently curving line. The spline, held in position by weights with hooks inserted in its slotted top, can be made to pass through specified key points. Flexibility is limited, however, and a tortuous curve must be drawn in sections, with care taken to avoid a noticeable change in direction at the junctions.

The spline curve process can be simulated by fitting a cubic polynomial of the form

$$Y = a + b_1X + b_2X^2 + b_3X^3$$

[5]Thomas K. Peucker and Nicholas Chrisman, "Cartographic Data Structures," *American Cartographer,* 2, no. 1 (April 1975), 55–69.

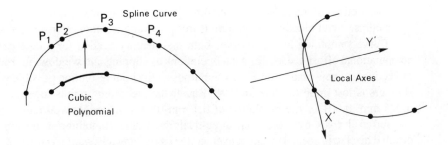

Figure 5-12 In cubic spline interpolation, each curved segment P_2P_3 is represented by a cubic polynominal curve based upon four successive points P_1, P_2, P_3 and P_4 (left). Local coordinate axes X' and Y' can be used to avoid steep slopes and multiple-valued functions (right).

to successive runs of four points (Fig. 5-12, left). The intercept and coefficients are estimated by the least-squares method used for polynomial trend surfaces. Use of an equation with only four parameters, a, b_1, b_2, and b_3, to represent the contour between only four points assures a perfect fit. Once the curve is represented by the cubic equation, the additional points can be found by evaluating the equation for Y at intermediate values of X. These intermediate points can be selected sufficiently close to each other to create the visual impression of a curved line. To assure curved sections that will blend into each other at the junctions, a separate cubic polynomial is computed for each original straight-line segment, for which the two end points serve as the second and third members of the run of four points.[6]

The polynomial equation used in cubic spline interpolation requires that Y be a single-valued function of X. That is, for each value of X there is only one value of Y, so that within the range of the four points used for curve fitting a line perpendicular to the X axis will encounter the spline curve only once. At times, of course, a contour line might reverse its direction, for example, by first trending upward to the right and then bending back toward the left. In some instances a simple transposition of X and Y, to yield

$$X = a + b_1Y + b_2Y^2 + b_3Y^3$$

will provide the necessary single-valued formula. In other cases rotation to a set of local axes, one of which passes through the second and third of the four points, might be needed (Fig. 5-12, right). Occasionally, as with four points defining a small, closed

---

[6]The curved cubic spline sections have second-order mathematical continuity at the junctions; that is, the two adjoining sections exhibit both the same slope and the same rate of change of slope. These conditions assure a smooth transition between sections. See Wolfgang K. Giloi, *Interactive Computer Graphics* (Englewood Cliffs, N.J.: Prentice-Hall, Inc., 1978), 129–34.

contour, sections of the curve spanning three points rather than four might be approximated by an ellipse, with blending of adjoining arcs providing a smooth transition at the joints.[7]

Curve smoothing also conserves both the memory needed for a data base and the computational effort needed for such operations as clipping and windowing, rotation, and projecting to a new map base. For most purposes, the high resolution possible with optical scanners is lost upon the tolerant eye. Rather than recording points to, say, the nearest 0.025 mm (0.001 in.), a resolution of 0.1 mm (0.004 in.) might well reduce processing and retrieval costs by more than three-quarters. Curve smoothing at the time a finely detailed display is needed might conveniently remove any awkward jerkiness in the plotted vectors. If the map is to be plotted at a smaller scale, even this smoothing might be unnecessary.

Similar smoothing procedures are useful for assuring that features crossing the edges of map sheets digitized separately not only meet but also exhibit no abrupt bends. In edge matching, some points, usually near the edge, must be shifted to provide a smooth transition, and a single polynomial curve is usually fit by least-squares methods to several points on both sides of the join line. Points near the edge are then relocated to lie along this smoothed, averaged path of the feature.

Label Placement

On many maps lettering is the principal type of symbol. Place names, for example, are used more frequently than meridians, parallels, and other grid lines to tie mapped information to parts of the environment familiar to the map viewer. For some thematic maps, such as land-use, geologic and soils maps, one-, two-, or three-letter codes can be more useful than graphic area symbols differing in value or pattern. Although graphic symbols might promote visual search if a location with known traits is sought, abbreviations can be ordered alphabetically and are thus located more easily in a complex legend by the map user who wants to know the characteristics of a particular location. The map user familiar with the map and data might easily decode, without repetitious reference to a legend, simple abbreviations, such as HnB representing the "Honeoye silt loam, with 2 to 8 percent slopes" on a soils map. With just the 26 letters of the alphabet, 17,576 (26^3) unique three-letter codes are available, many more than the largest number of visually differentiable, color-coded, patterned-area symbols that might possibly be designed, and surely a sufficient number to accommodate most classifications of geographic phenomena with good mnemonics.

In traditional cartographic drafting, the careful placement of labels is the bane of many beginners, but computer-controlled plotters can center labels with precision and maintain perfectly parallel alignments where needed as well. If properly programmed, a computer graphics system might also provide curved instead of straight bases for the

---

[7]Curve fitting can be quite complex when simple polynomials cannot be used. See Sylvan H. Chasen, *Geometric Principles and Procedures for Computer Graphic Applications* (Englewood Cliffs, N.J.: Prentice-Hall, Inc., 1978), 11–123.

names of rivers, so that labels might better conform to the shapes of their features. Curved labels can also be plotted to follow a parallel of latitude on a conic projection, with a result far more graceful than the bland, straight-line labels commonly produced with stick-up type. Because individual letters can be translated, rotated, and scaled to accommodate both large areas and small areas, map labels can easily be expanded or condensed, enlarged or reduced, as appropriate for individual situations.

Although computer-assisted cartography can produce labels geometrically precise in alignment and continuously variable in size, the computer has, as it were, a limited field of vision and might easily plot a label over a mapped feature or another label. One solution is the intervention of a human operator to make adjustments in label position after the display of an initial attempt. Another, more fully automated approach might have the computer select a tentative position for a label and then compute the distance to every previously plotted feature in order to prevent overlap. This expensive "brute-force" approach can be avoided if the data structure provides links identifying each feature's neighbors.

Because of the intricate shapes that must usually be contained within a small space, computer-produced lettering is particularly vulnerable to the graphic problems resulting from an irregular ink flow and an absorbent drawing surface. Graceful, tight curves require many short straight-line segments within a small area, slowing down plotting and possibly contributing to an uneven flow of ink. Consequently, most letters plotted in ink

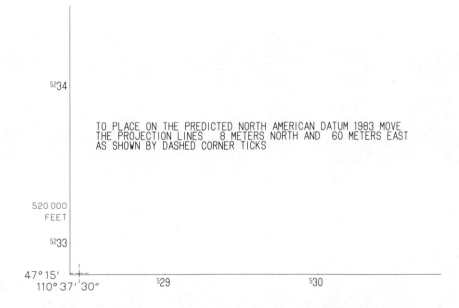

Figure 5-13 High graphic-arts quality labels and line work produced by film plotter at the U.S. Geological Survey for the collar of orthophotoquad sheets. (Courtesy U.S. Geological Survey.)

have abrupt rather than rounded curves. Characters with both wide and narrow lines require either pens with different line widths or a pen with a relatively thin width and multiple stroking for the wider lines. Multiple stroking can provide a wide variety of fonts, particularly for letters plotted at a relatively large size, say, taller than 18 mm (0.7 in.), and then photographically reduced.[8] In vector mode, very small letters, less than 3 mm (0.1 in.) tall, say, are better produced with a lighthead plotting on photographic film with either a small raster-mode CRT or a rotating turret mask with clear openings in the form of letters, numbers, and other special symbols (Fig. 5-13). Although expensive and not widely available, some graphic display devices can surpass the precision and esthetic quality of the most skilled cartographic draftsperson. Thus the challenge for computer-assisted cartography is not "can it be done," but "can it be done economically."

[8]Kurt E. Brassel and Jack J. Utano, "Font Variations in Vector Plotter-Lettering," *Computer Graphics,* 12, no. 1 (Spring 1978), 67–77.

6

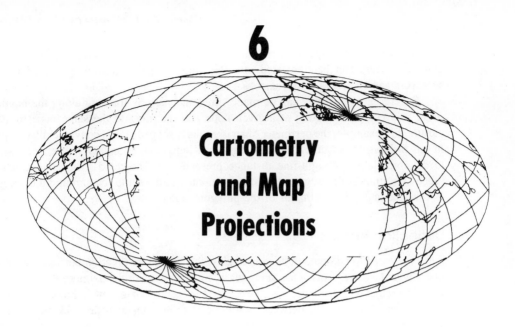

Cartometry and Map Projections

The fundamental difference between raster and vector data explains why most analytical operations can be carried out more efficiently in raster mode: raster data are organized for the ready retrieval of neighboring cells in any geographic direction, whereas vector data usually consist of sequential lists of coordinates without an implicit designation of neighbors. Nonetheless, because vector-mode lists frequently describe linear features or area polygons, the precise measurement of feature length or polygon area is particularly convenient. Although analyses requiring knowledge of a place's neighbors can be difficult, vector data are most amenable to measurement operations and geometric transformations involving either all coordinates in a file treated individually, as in map projection, or polygons and linear features treated separately, as in cartometry.

CARTOMETRY

Many applications of computer-assisted cartography consist of such relatively straight-forward operations as data capture and simple display, with the advantages of automation lying largely in the rapid generation of new maps after additions and corrections are made. A good example is the parcel block inventory that might be used by a land developer or survey engineering firm to plan a subdivision and account for individual lots. Computer-assisted mapping with this relatively rudimentary geographic information system has the useful side benefit of promoting the fast and accurate measurement of lot areas and the lengths of roads and utility lines. Measurement of angles between straight-line boundary sections also is useful, particularly so because plans developed with, say, an interactive graphics system can be translated into the precise legal descriptions needed by field surveyors and recorders of deeds.

Several measurements can be made simultaneously by tracking along the boundary of an area polygon, starting at a given node and returning to the same node to assure closure. For example, the perimeter of an irregularly shaped area is conveniently computed in this fashion, by accumulating the lengths of individual boundary segments according to the formula for straight-line distance presented in Chapter 5. The angles between successive straight-line segments can be computed in the same scan around the polygon, as can area and the coordinates of the polygon center.

Deflection Angles

Surveyors need to compute deflection angles, the change in direction between successive linear boundary segments. One approach is first to compute the azimuth $\theta_{(y)k}$, measured in degrees clockwise from the positive Y axis, for each line segment and then to determine the difference in azimuth between adjoining segments taken in a clockwise sequence around the polygon. For segment k running from (x_k, y_k) to (x_{k+1}, y_{k+1}), the tangent of azimuth $\theta_{(y)k}$ is computed as

$$\tan \theta_{(y)k} = \frac{x_{k+1} - x_k}{y_{k+1} - y_k}$$

from which the angle $\theta_{(y)k}$ is recovered by the arc tangent function (Fig. 6-1, left).[1] The bearing of segment $k + 1$ relative to segment k can then be computed as $\theta_{(y)k+1} - \theta_{(y)k}$ (Fig. 6-1, right). A negative deflection angle, indicating a counterclockwise shift in directional trend, can be converted to a positive, clockwise azimuth by adding 360 degrees.

Area

Polygon area can be computed by accumulating a running balance of triangular areas. Consider the polygon in Figure 6-2. Note the four nodes, the four sides, and the four triangles that are formed by the nodes bounding each side and the origin of the coordinate

---

[1] If y_{k+1} is very near y_k, this formula cannot be used because the resulting tangent would be close to infinity, too large a number for a digital computer. In this event, one of two possibilities must be recognized. If x_{k+1} exceeds x_k, $\theta_{(y)k}$ is 90 degrees, whereas if x_k exceeds x_{k+1}, $\theta_{(y)}k$ is 270 degrees.

The algorithm must make another, more common adjustment. Like other trigonometric functions, the tangent is periodic in the interval 0 to 360 degrees, with a basic wave completely defined between 90 and 270 degrees. Because built-in software functions usually evaluate the arc tangent function for the interval -90 to 90 degrees, 180 degrees must be added to the computed arc tangent if the tangent is positive and y_{k+1} is less than y_k. If the tangent is negative, the desired azimuth in degrees clockwise from the positive Y axis requires that the computed arc tangent be subtracted from 180 degrees if y_{k+1} is greater than y_k and from 360 degrees if y_{k+1} is less than y_k.

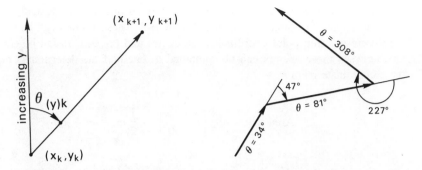

Figure 6-1 Azimuths relative to the positive Y direction can be computed for individual directed line segments (left) and used to measure relative azimuths or bearings as angular shifts in direction (right).

axes. These triangles overlap. Note that the area of polygon *ABCD* is equal to the areas of triangles *BCO* and *CDO* minus the areas of triangles *ABO* and *DAO*. Expressed differently, the area of the polygon is the sum of the triangular areas formed by the origin with sides 2 and 3 minus the triangular areas formed by the origin with sides 1 and 4.

The areas of these triangles can be computed as a variation of the *vector cross-*

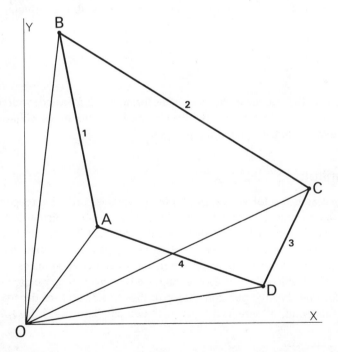

Figure 6-2 Four-sided polygon with four overlapping triangles between each side and the origin.

product.[2] As a demonstration, consider the directed line segments, also called vectors, from the origin O to points A and B in Figure 6-2. These vectors can be represented by their corresponding point coordinates as (x_a, y_a) and (x_b, y_b). It can be shown that the area between these vectors can be computed as one-half the determinant computed for these coordinate pairs as

$$\text{Area } ABO = \frac{1}{2} \begin{vmatrix} x_a & x_b \\ y_a & y_b \end{vmatrix} = \frac{1}{2}(x_a y_b - y_a x_b)$$

Note that the order of these two vectors in the determinant is important; reversing points A and B by exchanging columns would change the sign of the result. When the counterclockwise angle that would rotate vector A into alignment with vector B is less than 180 degrees, the triangular area is positive. Otherwise, the triangular area is negative. The sign, positive or negative, is significant because, for the example shown in Figure 6-2, the determinant formula yields positive areas for triangles ABO and DAO and negative areas for triangles BCO and CDO. The resulting area computed for polygon $ABCD$ is thus negative, because the nodes and triangles are encountered in a clockwise sequence, and would have to be multiplied by -1.

The calculation of area with positive and negative triangles can be generalized to a polygon of n nodes, taken in clockwise order, with the formula

$$\text{Area} = (-\tfrac{1}{2}) \sum_{i=1}^{n} (x_i y_{i+1} - y_i x_{i+1})$$

Note that when i assumes the value n, the formula calls for a node with coordinates (x_{n+1}, y_{n+1}); this final point has the coordinates (x_1, y_1) of the initial point, which is repeated to close the boundary around the polygon.

Centroids

Area centers, useful as locations for point symbols portraying properties of area polygons, can be computed as an adjunct of the polygon area calculation. A similar piecemeal approach is used, with each edge of the polygon contributing either positively or negatively to a running balance. These positive and negative contributions can be accumulated simultaneously for the polygon's area and its center coordinates.

The mathematical principle employed in calculating the area centroid is the *trapezoidal rule*. For the polygon shown in Figure 6-3, each of the five sides forms the top of a trapezoid, a four-sided figure with two parallel sides. Edge AB, for instance, is the

---

[2]See, for example, Merle C. Potter, *Mathematical Methods in the Physical Sciences* (Englewood Cliffs, N.J.: Prentice-Hall, Inc., 1978), pp. 204–207.

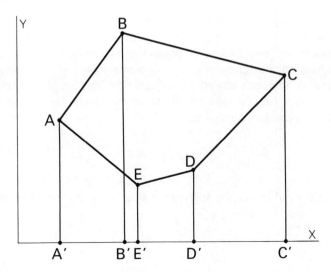

Figure 6-3 Five-sided polygon with five overlapping trapezoids used to compute x coordinate of area centroid.

top of trapezoid $ABB'A'$ formed by dropping lines perpendicular to the X axis from points A and B. Because the area centroid would also be the center of gravity of a polygon formed from a thin plate of uniformly dense metal, centroid coordinates can be computed as the weighted averages of the center coordinates for the polygon's trapezoidal components. Thus, with the polygon in Figure 6-3, the positive contributions of $ABB'A'$ and $BCC'B'$ and the negative contributions of $DCC'D'$, $EDD'E'$, and $AEE'A'$ can be accumulated for each trapezoid and used to compute the weighted averages

$$x_{\text{cent}} = \frac{\sum x_{\text{trapezoid}} A_{\text{trapezoid}}}{\sum A_{\text{trapezoid}}}$$

and

$$y_{\text{cent}} = \frac{\sum y_{\text{trapezoid}} A_{\text{trapezoid}}}{\sum A_{\text{trapezoid}}}$$

In these formulas the centerpoint coordinates of the individual trapezoids are weighted by the trapezoid areas. In the example, the trapezoids capped by polygon edges DC, ED, and AE should be given negative weights to counterbalance the parts of the other two trapezoids lying below the polygon.

Designation of area weights as positive or negative is automatic. As with the calculation of polygon area, direction around the polygon is important in determining centroid coordinates. For a clockwise path, positive contributions are entered for segments directed to the right, that is, with the second x coordinate x_{k+1} greater than the first x coordinate x_k.

In contrast, negative contributions result from segments directed to the left, with x_{k+1} less than x_k. The appropriate positive or negative weights can be obtained while computing the area of each trapezoid as the product of its average height and the length of its base. If the base length is given as $x_{i+1} - x_i$, where point $i + 1$ follows point i in a clockwise sequence of segments around the polygon, the base length will be negative for "lower" edges, such as *CD, DE,* and *EA*. When combined with the average y coordinate, the area of each trapezoid, computed as

$$\text{Trapezoid area} = (x_{i+1} - x_i) \frac{y_{i+1} + y_i}{2}$$

will yield weights with the appropriate sign. Thus the overall x center can be computed as the weighted average

$$x_{\text{cent}} = \frac{\sum\limits_{i=1}^{n} [(x_{i+1} + x_i)/2] \, (x_{i+1} - x_i) \, [(y_{i+1} + y_i)/2]}{\sum\limits_{i=1}^{n} (x_{i+1} - x_i) \, [(y_{i+1} + y_i)/2]}$$

The corresponding y center can be computed by a similar formula, but with x and y reversed, for trapezoids perpendicular to the Y axis, instead of to the X axis.

Other definitions of the center might be used. These include the centers of inscribed and circumscribed circles and the center of the smallest rectangle that might be drawn around the polygon. Because the area centroid of an irregularly shaped polygon might lie outside the polygon, the center of the inscribed circle or some other definition of the polygon center would be appropriate for locating point symbols within highly irregular areas.

Point-in-polygon Matching

Determining whether an area centroid falls within a polygon is but one application of point-in-polygon algorithms. Queries frequently are made to land-use information systems about the characteristics of point locations. With vector polygon data, matching areal units with included points can be a time-consuming task for the computer. Whereas with raster data the association of point coordinates with the corresponding data record is usually straightforward, with vector polygons the matching process is likely to involve numerous comparisons, possibly with each point tested against every areal unit in the data file until a polygon containing the point is encountered.

Computational complexity can, of course, be reduced by organizing the polygons in an areal hierarchy so that a match is attempted first with a region and then, once the region containing the point is found, only with polygons inside that region. For example, relating a point to one of the more than 3,000 counties in the United States can usually be accomplished in considerably less time by first matching the point with one of the four

Census regions, then with one of the states contained in that region, and finally with one of the counties within the state. Because several states have over 100 counties (Texas alone has 254 counties) further subdivision is useful on a selective basis. The resulting *tree structure* of polygons nested in this manner is discussed in Chapter 7.

Matching time can be reduced further if the coordinates of the smallest rectangular window bounding each polygon are included in the data base. In this way, extensive but fruitless checking can be avoided when a preliminary inspection reveals that the point is above, below, to the left of, or to the right of this window.

After a successful match with a polygon's bounding window, further checking is necessary, of course, because a point within this window need not be within the polygon. The simplest and most straightforward test for a point-in-polygon match is the *plumb-line algorithm,* the equivalent of dropping a line directly downward from the point and counting the number of times, if any, this line crosses the polygon boundary.[3] If an odd number of intersections occurs between the polygon and the plumb line, the point is within the polygon; otherwise, the point lies outside and the search for a match is unsuccessful and must continue (Fig. 6-4, left).

The plumb-line algorithm is similar in principle to the procedure described in Chapter 5 for shading area polygons. Yet, because the plumb line is parallel to the Y axis, the calculations are simpler. As with other polygon calculations, each edge must be examined separately. Because an intersection with the plumb line will occur only if the edge in question is below the point, further testing of the edge is not needed if the y coordinates of both ends are greater than the y coordinate of the point (Fig. 6-4, right). Furthermore, an intersection can occur only when the x coordinate of the point is within the range of x coordinates of the ends of the edge. Thus, if the point is within the x coordinate range of the segment and above both end points, an intersection can be assumed and need not be calculated. Should one end of the edge lie above the point and the other end lie below, the y coordinate of the intersection must be computed and examined so that only intersections between polygon and plumb line are counted.[4] If the maximum and minimum x and y coordinates are stored for each polygon, further time is saved because many polygons can be dismissed readily as too far to the left or right, too high, or too low.

Overlay Analysis

Although point-in-polygon search can be made efficient for sampling area polygon files by choosing a suitable data structure, vector-mode processing can be quite cumbersome for the overlay analyses discussed in Chapter 4 for raster data. With the simple search

---

[3]For a different approach, based upon angles between lines from the point to the polygon nodes, see Stig Nordbeck and Bengt Rystedt, "Computer Cartography: Point in Polygon Programs," *Lund Studies in Geography,* Ser. C, no. 7, 1967.

[4]Double counting of intersections at polygon nodes can be avoided by assigning each edge only one end point and comparing the x coordinate of the point with the coordinate range for the edge with the test $x_i > x_{point} \geq x_{i+1}$.

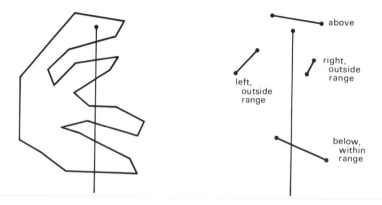

Figure 6-4 A point is within a polygon if its plumb line has an odd number of boundary intersections (left), as might be determined by comparing the point's coordinates with the coordinate ranges of individual boundary segments (right).

for plumb-line intersections with polygon edges, computational effort is directly proportional to the number of boundary segments. In contrast, an attempt to find the intersections of boundaries for two polygons can require an effort proportional to the products of the numbers of segments in the two polygons. Hence, maintaining separate polygons for every bounded area and a separate file of polygons for every characteristic of interest is likely to be costly and slow.

A vector-mode approach that obviates the calculation of polygon intersections can provide a computationally efficient solution to overlay analysis. If every place in the region covered by the inventory is assigned to one and only one polygon, intersections between polygons need not be computed. Each polygon would have an attribute list that could be inspected for the desired combination of traits. With this arrangement of the data, the intersection and union operators of set theory do not require a search for intersecting boundaries. Only when the analysis is restricted to an irregularly shaped subregion or when a rectangular window is to be displayed must boundary intersections be computed.

Although this data structure can simplify overlay analysis with vector polygons, graphic display and further processing are usually simplified if boundaries are eliminated between adjoining areal units both having the same specified properties. Once again, an appropriate organization of the data can reduce computational effort. If boundaries are stored as groups of linked segments between nodes at which three or more boundaries meet, each polygon can be represented as a relatively simple list of these segment groups or *chains* taken in a clockwise sense around the perimeter. The information maintained for each chain of boundary segments should identify the polygons lying to the left and right. With this information available, the chain lists for two adjoining polygons can be merged and the common boundary chains dropped (Fig. 6-5). Additional polygons abutting either of these two polygons can join the cluster in a similar manner, one at a time.

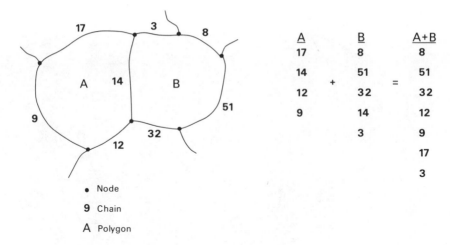

A	B	A+B
17	8	8
14	51	51
12	32	32
9	14	12
	3	9
		17
		3

(A + B = A+B)

• Node
9 Chain
A Polygon

Figure 6-5 Polygons A and B, with boundary coordinates organized as segment chains between nodes (left), can be combined by merging the segment lists and dropping common segment chains (right).

PROJECTIONS AND TRANSFORMATIONS

Map projections have two fundamental uses: problem solving and communication. The Mercator and gnomonic projections, on which straight lines are, respectively, lines of constant geographic direction and great circle routes, have been indispensable for marine and air navigation. Yet both projections greatly distort relative area and thus are of little use for communicating geographic facts not related to navigation or direction. Nonetheless, these projections, perhaps because they have been readily available as base maps, have been used in totally inappropriate situations, particularly in the case of the Mercator, which once was the wall map of the world in many elementary schools. These blatant mismatches between projection and map objective would appear to be relatively rare today, but many maps are still made with projections far less effective than others that might be employed had the cartographer used more initiative. Computer-assisted cartography, by displacing the "let's see what base maps we have here in the drawer" approach to selecting map projections, can provide solutions that are effective as well as convenient.

Conventional Projections

Most map projections are easily plotted in vector mode. Elaborate data structures generally are not needed because the calculations usually involve the straightforward evaluation of two mathematical formulas, one for the x coordinate and the other for the y coordinate, transforming latitude and longitude to plane coordinates. Each boundary, coast, or feature line is transformed and plotted, point by point.

Most world map projections are accompanied by their graticule, a regular grid of

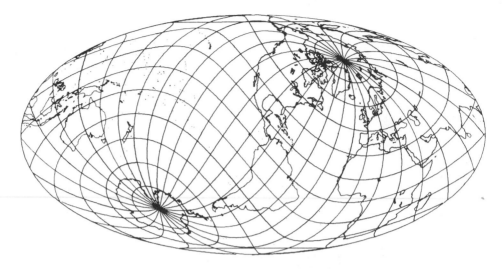

Figure 6-6 Oblique Mollweide projection, centered near the Galapagos Islands. Persons with an outline data base such as World Data Bank I or II and handy with programming can easily create customized map projections. Mapping programs such as CAM can be used to draw specially centered versions of many standard projections. (Courtesy University·of Wisconsin Cartographic Laboratory.)

parallels and meridians. Points along these grid lines are usually generated by the program, with the increments in latitude and longitude chosen by the program user. Calculations can often be simplified if the programmer recognizes such straight lines as the meridians on polar azimuthal and normal conic projections; only the two end points need be calculated. Moreover, on many projections meridians or parallels are the arcs of circles for which the projection equations need only be consulted once to determine the end points and radii. Yet, in plotting grid lines with a varying curvature, for which individual points must be generated from the projection formulas, the interval between successive points can be computed to yield a smaller density of points along the more tightly curved, more visually sensitive portions of grid lines.[5]

Map projections distort scale systematically, usually with a point, a line, or a pair of lines serving as an anchor or framework away from which distortion increases. With computer-assisted cartography, the map maker focusing on a particular part of the world is able to center the projection on that point, in contrast to most outline maps, which are centered commonly on the equator or one of the poles (Fig. 6-6). In addition to choosing those parts of the map with the least distortion, the cartographer has several other decisions, particularly the scale, the overall appearance of the Earth's grid, and whether to preserve

[5]For a fuller discussion of the role of the computer in the generation and plotting of map projections, see Waldo R. Tobler, "Numerical Approaches to Map Projections," in Ingrid Kretschmer, ed., *Studies in Theoretical Cartography: Festschrift für Erik Arnberger* (Vienna: Franz Deuticke, 1977), pp. 51–64.

angles or relative area; a projection cannot be both conformal and equivalent, but might be neither. Distortion diagrams, with isolines showing the pattern of angular or areal distortion or with point symbols portraying relative stretching and compression, can also be plotted to aid in these decisions and to produce training aids for cartographers and map users who want to understand more fully the nature of map projections.[6]

Cartograms

The distortion that is inevitable in map projections is manipulated in the cartogram to serve specific objectives in map communication. Two principal types of cartograms are used: the linear cartogram, which distorts distance to represent such distance-related concepts as travel time and transport cost, and the area cartogram, which adjusts the mapped size of each areal unit in proportion to its relative importance, for example, in population or industrial production.

Like the azimuthal equidistant projection, the linear cartogram has a single focus, an origin or a destination. Because the symbols usually are limited to labeled dots representing places and labeled concentric circles representing time or cost, many cartographers consider this type of map more a statistical diagram than a map projection. This distinction is inappropriate, however stark and unmaplike linear cartograms may appear, because mathematical transformations can be found so that additional features, boundaries, rivers, routes, and even topographic contours can be "carried along" as transport time or cost is allowed to distort the map.[7]

Area cartograms provide a dramatic picture of the disparity between land area and the values at areal units of a transforming variable. News magazines, as a result, have used cartograms to emphasize the importance in presidential elections of small but populous states with many electoral votes. Because map viewers might otherwise associate the area covered by a color symbol representing states likely to support a particular candidate with the candidate's probable success, the area cartogram can be useful as an electoral base map. More generally, the cartogram can provide a demographic base map when, for example, the map author chooses to remind the map viewer that population density is not uniform or to counteract tendencies of area symbols to command attention in proportion to the area covered on the map. The demographic base map thus can be a useful tool for presenting mortality rates so that the map reader will not be overly impressed by large but sparsely inhabited areas with extremely high or low death rates.

There are two varieties of area cartograms, contiguous and noncontiguous. Contiguous-area cartograms, which can be constructed automatically, retain all boundary relationships without any tearing or separations, and consequently might severely stretch and contort shapes in regions where land area is inversely related to the value of the

---

[6]Peter Richardus and Ron K. Adler, *Map Projections for Geodesists, Cartographers and Geographers* (Amsterdam, London and New York: North-Holland Publishing Company and American Elsevier Publishing Co., Inc., 1972), pp. 129–35.

[7]See, for example, W. R. Tobler, *Bidimensional Regression: A Computer Program* (Santa Barbara, Calif.: Geography Department, University of California, Santa Barbara, 1977).

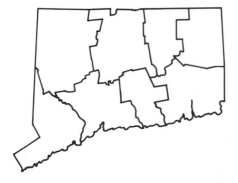
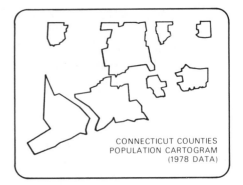

CONNECTICUT COUNTIES
POPULATION CARTOGRAM
(1978 DATA)

Figure 6-7 Noncontiguous area cartogram (right) obtained for Connecticut counties (left) by scaling areas in proportion to population and then rearranging the area outlines on an interactive graphics system to avoid overlap. (Courtesy Anthony V. Williams.)

transforming variable.[8] Noncontiguous-area cartograms merely scale the outlines of areal units so that map area corresponds to such magnitudes as population or agricultural acreage. Place identification is promoted because shapes are not further distorted, but contiguity is lost as boundaries contract inward toward the centers of each areal outline.

In a sense, a noncontiguous-area cartogram is a map using graduated point symbols each of which portrays the boundaries of its corresponding areal unit (Fig. 6-7). Thus each area polygon can be developed by computing its area centroid and internal area, translating the polygon to a temporary center at the origin for scaling, and translating the scaled outline back to its original center for plotting.[9] Overlapping might be avoided by choosing a scale factor that will reduce in size all areal outlines except that of the area with the greatest density of the mapped phenomenon. The appropriate linear scale factor s_i for an areal unit with area A_i and data value V_i for the transforming variable is

$$s_i = \left(\frac{cV_i}{A_i} \right)^{1/2}$$

where scaling constant c is the inverse of the largest density, that is, $1/(V/A)_{max}$. Because irregularly shaped, "interlocking" areal units might contract across boundaries into their neighbors when scaled downward, an interactive graphics system is useful for repositioning area centers after viewing an initial plot. Visually determined centers in some cases are preferred to mathematically defined ones.

---

[8]See Waldo R. Tobler, "Geographic Area and Map Projections," *Geographical Review*, 53, no. 1 (January 1963), 59–78.

[9]Judy M. Olson, "Noncontiguous Area Cartograms," *Professional Geographer*, 18, no. 4 (November 1976), 371–80.

Reprojection for Clarity

Map projections can be developed to solve many problems in cartographic communication, including the lessening of clutter caused by densely packed, possibly overlapping symbols in some parts of the map. A common example of a region cluttered with symbols is the United States' Northeast Megalopolis, stretching from Portland, Maine, southward to the Hampton Roads area in Virginia. Here many otherwise important features often must be suppressed to avoid a visually congested and confusing mass of symbols. In some cases this congestion could be avoided by increasing the scale of the map, but a uniform enlargement might well result in a map too large for a given format, such as a page in a book, periodical, or report. An inset map might be used, to set apart the congested region for portrayal at a larger scale, but the resulting interruption in the mapped pattern also can be undesirable for some presentations. Because distortion of some type is inevitable on small-scale maps, why not mold the projection to avoid clutter and provide for more detail in areas with greater densities of features? Given the proper cues by familiar boundaries and place names, map viewers can, after all, recognize shapes and identify places on other systematically distorted projections, such as the Mercator and the gnomonic.

For a fuller understanding of reprojection, take an outline map of the United States (you should be able to do this mentally if you are at all familiar with the geography of North America) and hold it in front of you with the east coast to the right. Bend the left part of the map farther away from you and note that the apparent scale of the map is now larger in the east than in the west. Note as well that the apparent horizontal size of the map has decreased, and that the resulting projection might be suitable for a single-column map in a newspaper, particularly if named locations are to be plotted (Fig. 6-8, left). If east-west scale is not compressed anywhere on the map to less than one-fourth north-south scale, most people should be able to identify familiar areal shapes, even without the additional cues provided by relative location.[10]

In the reproduction process a simple transformation often need be applied to only one of the two coordinates, x and y, obtained at an earlier stage by a map projection from the globe to a plane. For example, the linear transformation of the x coordinate

$$X' = 0.7X$$

will compress in the east-west direction a map, with north at the top, in the direction of the positive Y axis. In this case the new east-west scale is uniformly adjusted to be 70 percent of the unreprojected east-west scale. The more selective distortion described earlier for the U.S. outline map in order to avoid overlapping symbols in the East would require a slightly more complex, nonlinear function, such as that plotted in the right half of Figure 6-8. In addition to these one-dimensional transformations, simple reprojection can also be point centered, as when scale on the map of an urban area is made to vary

---

[10]Mark S. Monmonier, "Nonlinear Reprojection to Reduce the Congestion of Symbols on Thematic Maps," *Canadian Cartographer*, 14, no. 1 (June 1977), 35–47.

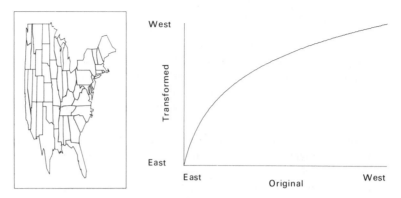

Figure 6-8 Reprojected outline map of the conterminous United States (left) based upon a logarithmic transformation (right) of east-west scale. (Source: Mark S. Monmonier, "Nonlinear Reprojection to Reduce the Congestion of Symbols on Thematic Maps," *The Canadian Cartographer,* 14, no. 1 [June 1977], 35–47. Reprinted by permission of the publisher.)

inversely with distance from the center of the business district.[11] Although by no means new, variable-scale maps can now be generated with relative ease by computer-assisted cartography.[12]

STEREOSCOPIC VIEWS

Many maps require the graphic representation of height for socioeconomic phenomena such as income and population, as well as for physical phenomena such as elevation and barometric pressure. Standard approaches to this task include isolines on topographic maps and various graytones on choropleth maps. Symbolization in the traditional sense of the term is not the only means of representing height differences, for the assignment of locations to map symbols, the traditional role of map projections, can enhance dramatically the viewer's perception of surface elevation and form. Specifically, if a different map projection is prepared for each eye, the viewer's stereoscopic vision can be made to "see" mapped heights even though the map is flat, rather than three dimensional.

Two simple approaches can be used to convince the brain that each eye is viewing the same three-dimensional scene from its own, slightly different vantage point. The *anaglyph* uses two superimposed maps, each representing the same surface from a slightly different view, and each plotted in a different color, usually red or green. The viewer wears a pair of filter glasses that transmit only reflected red light to the left eye and reflected green light to the right eye. Whereas the two images separated by filters in the

[11]Naftali Kadmon, "Data-bank Derived Hyperbolic-scale Equitemporal Town Maps," *International Yearbook of Cartography,* 15 (1975), 47–54.

[12]Waldo R. Tobler, "Local Map Projections," *American Cartographer,* 1, no. 1 (April 1974), 51–62.

anaglyph process share the same area on the paper, with the *stereo map* two physically separated but adjacent images are plotted, one for each eye. Instead of focusing on the paper, as for an anaglyph representation, the eyes viewing a stereo map must look directly ahead, toward infinity, so that the two lines of sight are parallel and thus able to visually fix corresponding elements on the separate views. Although the individual stereo images are limited in size by the average separation, approximately 65 mm, of the viewer's eyes, a more highly detailed than normal map can be printed by photographically reducing a pair of larger drawings for viewing under magnification with a pocket lens stereoscope used for aerial photograph interpretation.

The geometry of the anaglyph map is based on a simple principle: to make an image appear to lie above the paper, shift it away from the position where the eye would find the unelevated image to the intersection with the paper of a line of sight through the elevated point. Thus, if elevated point P in Figure 6-9 is plotted at P_r for the right eye and at P_l for the left eye, the eye-brain system, in fixating each eye on the appropriate filtered image, will perceive the elevation as z. The appropriate offsets to points where lines of sight intersect the paper are computed as

$$s_r = \frac{zx_r}{w - z}$$

and

$$s_l = \frac{zx_l}{w - z}$$

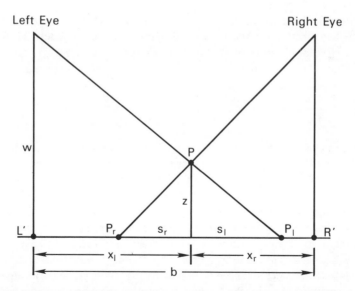

Figure 6-9 Geometry of stereo viewing, showing the offsets s_r and s_l required in anaglyph maps viewed at distance w to effect an apparent elevation z of point P above the paper.

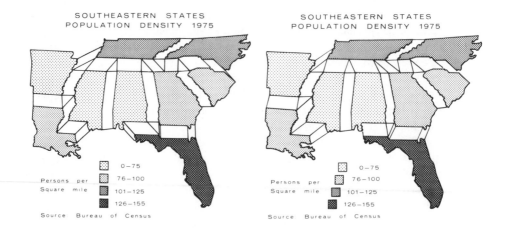

Figure 6-10 A stereoscopic pair of choropleth maps, with areal units offset to permit clear vision of tops of prisms. (Source: John R. Jensen, "Three-dimensional Choropleth Maps: Development and Aspects of Cartographic Communication," *The Canadian Cartographer,* 15, no. 2 [December 1978], 123–41. Reprinted by permission of the author and publisher.)

for a point located distances x_r and x_l from hypothetical eye centers R' and L' located arbitrarily near the center of the drawing, and where z is the intended elevation above the horizontal plane and viewing distance w is approximately 30 to 40 cm.

The designation of R' and L' in Figure 6-9 is arbitrary, because the sum of these offsets s_r and s_l, called the *parallax difference* and computed as

$$\Delta p = \frac{zb}{w - z}$$

where eye base b is the sum of x_r and x_l, is independent of the position of P relative to R' and L'. It is the parallax difference, not s_r and s_l individually, that convinces the eye and brain that point P lies distance z above the base plane.

This identical parallax difference can be communicated with a stereo map composed of the original, unaltered map and its *stereomate*. If the left eye views the untransformed map and the right eye views the stereomate, the appropriate parallax difference can be induced by shifting each point on the right-hand image distance Δp farther to the right. Thus a point plotted at (x, y) for the left image would have its right-hand counterpart plotted for the stereomate at $(x + b + \Delta p, y)$. If the plotted drawing is to be photographed and reduced in scale, the finished stereo map must be planned to provide an appropriate eye base b and the desired amount of vertical exaggeration.

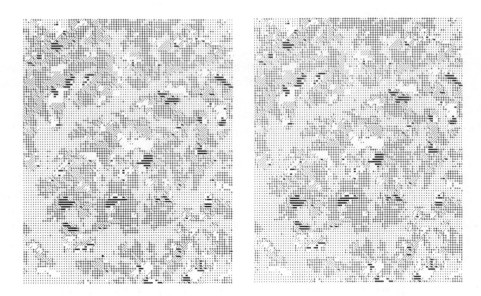

Figure 6-11 A stereoscopic pair of statistical surfaces with z-values portraying the rate of evapotranspiration and the pixels coded according to land cover: water (-), vegetation (•), corn (+), peanuts (thick horizontal lines), soybeans (/), and barren/urban (blank). (Courtesy John R. Jensen.)

A variety of three-dimensional symbols can be portrayed effectively with anaglyph and stereo maps. Contour lines, for example, can appear to float above the paper at levels appropriate to the elevations represented. Similarly, the effectiveness of a choropleth map can be enhanced by having each areal unit appear to rise above the paper in accord with either its actual surface elevation for the mapped variable or, for classed data, the mean value for its category. In order to prevent a tall prism from blocking part of an adjacent lower prism, the outlines of both areal units should be shifted apart slightly (Fig. 6-10).[13] Moreover, a second variable can be introduced to a map, even for a qualitative variable such as land cover, by adding a stereomate with parallax based upon topographic elevation or some other relevant factor. Figure 6-11, for instance, uses qualitative vector point symbols to represent land-cover categories for grid cells; the addition of parallax based on the rate of evapotranspiration shows that areas covered by water (–) and natural vegetation (the dot pattern) have the highest rates of water loss. Among other map symbols that might be made more effective by the addition of a stereomate are place labels, proportional point symbols, and route flow lines.

[13]John R. Jensen, "Three-dimensional Choropleth Maps: Development and Aspects of Cartographic Communication," *Canadian Cartographer*, 15, no. 2 (December 1978), 123–41.

OBLIQUE VIEWS

Dramatic cartographic representations of the third dimension require neither the color reproduction needed for anaglyph maps nor the small size of stereo maps needed to accommodate the eye base. An oblique view of a surface, plotted as the projections onto a picture plane of a series of vertical or inclined profiles, provides a visually effective representation of slopes, peaks, and pits (Fig. 6-12). The visual impression of the third dimension usually is enhanced by the removal of all lines that would not be visible if the surface were opaque. Thus oblique views are usually incomplete because some depressions, troughs, and backslopes are obscured by higher features lying closer to the picture plane. The oblique view sacrifices the more complete two-dimensional overview permitted by contour lines for an improvement in the perception of the third, vertical dimension.

Oblique views, plotted with vector symbols, are readily generated from raster data.[14] If necessary, interpolated surface values first are estimated for grid intersections from irregularly spaced data points. For the widely used SYMVU program, an oblique surface package distributed by Harvard University's Laboratory for Computer Graphics and Spatial Analysis, this initial stage of processing was relegated to SYMAP, the Laboratory's line-printer program for surface mapping. With the SYMAP-SYMVU combination and other surface viewing programs, such as the Laboratory's more recent PRISM package,

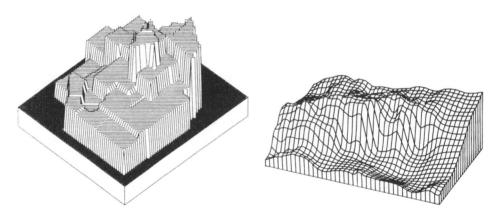

Figure 6-12 Examples of oblique surfaces with profiles along grid diagonals (SYMVU, at left) and rows and columns (SURFACE II, at right). (Source: *figure on right*—Robert J. Sampson, *SURFACE II Graphics System* [Lawrence, Kansas: Kansas Geological Survey, 1975], p. 221. Reprinted by permission of the publisher.)

---

[14]Vector polygon data can be used directly without conversion to a grid. See, for example, W. R. Tobler, "A Computer Program to Draw Perspective Views of Geographical Data," *American Cartographer*, 1, no. 2 (October 1974), 124, 190.

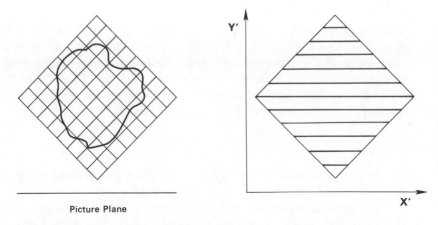

Figure 6-13 Orthogonal profile directions for block-base grid with irregularly shaped mapped region (left) and new reference axes for diagonal profiles (right).

oblique views can be obtained for discontinuous choropleth and proximal surfaces, both examples of a "piecewise" continuous surface, as well as for the continuous surfaces usually portrayed with isolines.

Grid orientation should be chosen carefully, because perception of the third dimension is enhanced by a block-diagram base with one corner pointing directly toward the picture plane. The rectangular outline of the block base usually corresponds to the boundaries of the rectangular grid (Fig. 6-13, left). Interpolation to the grid is thus best deferred until a viewing direction is chosen.[15] For irregularly shaped regions the mapped area usually is surrounded by a noticeably broader base of zero or low values.

Profiles can be developed along both grid directions to produce the appropriately called *fishnet map* (Fig. 6-12, right), or only along diagonals through the grid parallel to the picture plane (Fig. 6-12, left). Generally, a sparse data grid, say, 20 by 20, is better represented by a coarse fishnet map, whereas a dense grid can be portrayed more gracefully with profiles only along diagonal lines parallel to the imaging plane.

Each diagonal profile is a list of (x, y, z) coordinates, where x and y are coordinates for the grid and z is surface elevation. Because all profiles lie in parallel planes, the calculations for the projections of the profiles onto the picture plane can be simplified by a rotation of the grid in the horizontal plane to new reference axes, X' parallel to the picture plane and Y' perpendicular to the picture plane (Fig. 6-13, right). Thus, the y' coordinate is a constant for each profile.

[15]Mark S. Monmonier, "Viewing Azimuth and Map Clarity," *Annals of the Association of American Geographers*, 68, no. 2 (June 1978), 180–95.

Oblique Projections

A variety of oblique transformations can be used to project the profiles onto the picture plane. One of the simplest and most useful is the isometric projection, more appropriately called the parallel or orthographic projection because the image of each point is specified by a perpendicular line to the picture plane from the point. Where U and V are, respectively, the horizontal and vertical axes on the picture plane, the orthographic view is specified by a direct correspondence between X' and U,

$$U = X'$$

and the generation of vertical image positions along axis V based solely upon surface elevation (Z) and distance from the picture plane (Y').

The relative contributions of a profile point's z and y' coordinates are determined by the elevation angle ϕ of the viewing position above the horizontal plane (Fig. 6-14). A point with zero elevation on a profile lying distance y' from the base of the picture plane will project onto the picture plane at distance $y' \sin \phi$ above the origin. A point at the same horizontal location but with nonzero elevation z will project distance $z \cos \phi$ farther upward on the picture plane. Thus the vertical projected coordinate is computed as

$$v = y' \sin \phi + z \cos \phi$$

This equation for vertical profiles is less complex than the transformation required for the less common inclined profile approach.[16] The elevation z may also be scaled vertically before projection.

Another common transformation of surface elevations to a two-dimensional picture plane is the central perspective projection. Image points on the picture plane are fixed by a "line of sight" from each point (x', y', z) to a central focus, or viewpoint, on the opposite side of the picture plane (Fig. 6-15). For a viewpoint or "eye" lying distances z_e above the horizontal plane and y_e behind the picture plane, the vertical coordinate of the image can be computed by apportioning the increase in elevation along the line of sight to portions y_e and y' of the distance between image and object points; thus

$$v = z + (z_e - z) \left(\frac{y'}{y_e + y'} \right)$$

The horizontal coordinate of the image point also is affected by the distance y' from the picture plane, with more distant features pulled inward toward the center to reduce their

---

[16]See, for example, Pinhas Yoeli, "Computer-aided Relief Presentation by Traces of Inclined Planes," *American Cartographer*, 3, no. 1 (April 1976), 75–85.

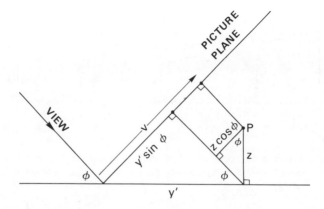

Figure 6-14 Vertical position V of orthogonally projected image of point P onto picture plane has components related to distance y' from base of picture plane and elevation z.

apparent scale on the picture plane.[17] This *foreshortening* is provided by the transformation

$$u = x' + (x_e - x') \left(\frac{y'}{y_e + y'} \right)$$

where x_e is the x' coordinate of the viewpoint (Fig. 6-15, right).

Other projections can be used, including a fisheye distortion and a pair of block diagrams produced with slightly offset viewpoints for stereo viewing. For most cartographic applications, however, the oblique orthographic view is generally satisfactory, particularly if the likely distance between the viewer and the reproduced map is unknown or variable, as when the image is to be projected from a transparency onto a screen to illustrate an oral presentation. More important concerns for the map designer are the selection of a meaningful combination of viewing direction, elevation angle, and vertical scaling factor. All three view parameters jointly control the visibility and informativeness of the oblique view, but the viewing direction usually is the more critical factor. Generally, an elevation angle of approximately 45 degrees is effective if the vertical scaling is chosen to make the surface appear neither overwhelmingly tall nor uninformatively low. An interactive graphics system can be most helpful in selecting an appropriate set of viewing parameters.

---

[17]This and other oblique views can be specified with a 4 by 4 transformation matrix using homogeneous coordinates, as discussed in Chapter 5 for the two-dimensional case; see Wolfgang K. Giloi, *Interactive Computer Graphics: Data Structures, Algorithms, Languages* (Englewood Cliffs, N.J.: Prentice-Hall, Inc., 1978), pp. 97–107.

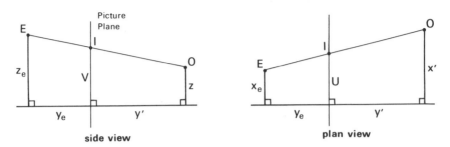

Figure 6-15 Side view (left) and plan view (right) of central projection of object point 0 onto picture plane at image point *I* with line of sight from viewpoint *E*.

Hidden Line Removal

The hidden surface problem in computer graphics has interested many researchers, and numerous algorithms have been developed to provide low-cost, low-memory solutions for this frequently encountered task. A relatively simple procedure used in many surface mapping programs is the scan-line method developed for surfaces represented by vertical profiles. Successive profiles are plotted, starting with that profile closest to the picture plane and proceeding toward the far end of surface (Fig. 6-16). In general, each profile will appear higher on the picture plane than the preceding profile unless an elevation farther removed from a previous, higher profile is sufficiently lower to be hidden. When hiding occurs, a profile must be disconnected and only the visible portion plotted.

The scan-line algorithm for hidden line removal operates by generating, at each step on the surface backward from the picture plane, a *current highest profile* consisting of all visible portions of the current profile and any parts of previous profiles that block from view portions of the present profile (Fig. 6-16). The height of the image at each point on the next profile is then compared with the elevation of the current highest profile at the same horizontal position. If a point on the new profile has a higher projected elevation on the picture plane, the point is visible and must be plotted. If the new point would project below the current highest profile, the point may not be plotted. Moreover, intersections between the new and current highest profiles must be computed so that only visible portions of the new profile are added to the map.[18]

Other solutions to the hidden-surface problem are used in computer-assisted cartography.[19] A minor modification of the hidden-line solution just discussed accommodates

---

[18]Because there is no foreshortening, these calculations generally are simpler for the orthographic than for the central projection. The test for hiding is complicated in the central projection because horizontal coordinates are not a simple function of grid position and the above-below test almost always requires comparing each point in the new profile with a line rather than with another point.

[19]In computer graphics a distinction often is made between the hidden-surface problem for pictures to be displayed on a raster display device and the hidden-line problem for pictures to be displayed as line drawings on vector display devices.

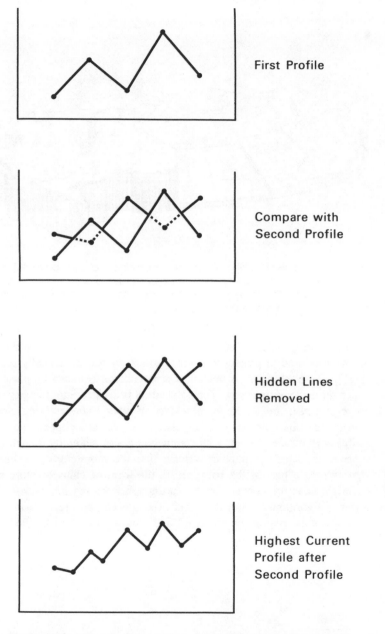

First Profile

Compare with
Second Profile

Hidden Lines
Removed

Highest Current
Profile after
Second Profile

Figure 6-16 Stages in the scan-line algorithm for hidden-line removal.

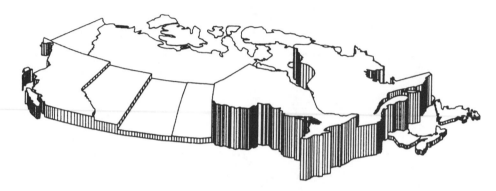

Figure 6-17 Oblique stepped-surface map plotted by PRISM, a program developed at the Laboratory for Computer Graphics and Spatial Analysis, Harvard University. (Courtesy Laboratory for Computer Graphics and Spatial Analysis, Harvard University.)

the two orthogonal sets of profiles used for fishnet maps.[20] A significantly different approach is used to project three-dimensional views of vertically scaled area polygons (Fig. 6-17). Each prism is treated as a polyhedron bounded by planes and the edges at which these planes intersect. The polyhedron is divided into triangles, which, in the very roughest sense, are tested for blocking by other triangles lying closer to the picture plane.[21] Invisible portions of triangular edges are then clipped and eliminated, and the visible portions are plotted, with common edges of adjoining triangles in the same plane suppressed. Computer graphics systems often incorporate this or a similar hidden-surface algorithm as a part of the hardware in the form of chips or other read-only memory (ROMs). Thus the concern of the cartographer for the intricacies of many computer-graphics algorithms eventually might need proceed no deeper than current, hardly recognizable concerns for the efficiency and accuracy of algorithms to evaluate the cosines of angles.

---

[20]See B. F. Sprunt, "Hidden-line Removal from Three-dimensional Maps and Diagrams," in John C. Davis and Michael J. McCullagh, eds., *Display and Analysis of Spatial Data* (New York: John Wiley & Sons, Inc., 1975), pp. 198–209.

[21]See, for example, Tobler, "A Computer Program to Draw Perspective Views."

7

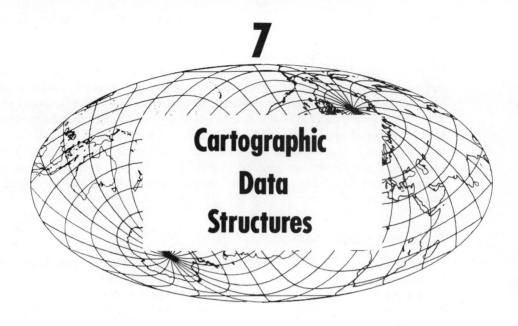

Cartographic Data Structures

Computational efficiency depends in large measure upon the organization, or structuring, of the data. In computers with virtual memory, for instance, it is important that data values likely to be accessed near each other in time be stored near each other in memory; otherwise, page thrashing is likely to impede processing, with overly frequent exchanges of data between internal and external memories. The general concern for efficient data structuring in information systems is particularly crucial for geographic information systems, for which the organization of the data often must reflect the need to search the data spatially, over two dimensions in a variety of directions, not unlike the way in which a person scans a printed map. Because large geographic data-handling systems can require the commitment of millions of dollars of public funds and many thousands of hours of human resources, and because these systems can influence public policy, the appropriations of many millions of dollars, and the future quality of the environment, their design must be approached carefully and cautiously, with a concern for future flexibility, as well as for present efficiency.

BASIC CONCEPTS

In computer science the study of data structures is far broader than the FORTRAN programmer's concern for, say, using one three-dimensional array or several two-dimensional arrays for a particular task. In practice as well as in principle, the efficiency and even the feasibility of computational processes relate not only to the algorithm and the program but also to the architecture of the computing system. Efficiencies achieved by the careful structuring of an algorithm are much less significant than efficiencies derived from the conscientious organization of a computer's memory. Computer architectures efficient for multiuser time-sharing, interactive graphics, and batch-processing scientific computing differ in as significant a way as the building architectures most suited for high school education, automobile production, and the treatment of mental patients. Both homes and computers can be planned to "save steps," and the agency or firm developing a geographic data-handling system should be as concerned with machine architecture as a family buying a new house is concerned with the number, types, and relative locations of rooms.

Addressing and Memory

Basic to many considerations in programming and computer design are the methods used for addressing memory locations. An important distinction is between the *computed address* and the *link* or *pointer*. Array element A(2, 3) in a FORTRAN program requires the calculation of an address from the values of the subscripts and the declared, or dimensioned, size of the array. If the array has 4 rows and 5 columns and is stored column by column, the element in the second row and third column occupies the tenth position in a part of memory labeled A, because it follows the 4 positions for each of the first two columns and the first-row element in the third column. Thus, if A begins at address 30124, A(2, 3) resides at 30133.

Addressing with links requires that each fundamental data object, for example, the pair of coordinates for a point along a boundary, be represented by a contiguous group, or *block,* of memory locations. In addition to memory cells containing the attributes of the data object, one or more locations in the block contain addresses pointing to the location of the "next" data object in memory. Thus three memory elements can represent each point along a linear feature, one location each for the x and y coordinates and the third for the address in memory of the next point along the line. Variable-sized data objects such as lines likely to be edited interactively are easily accommodated by this type of *list structure.* Data structures using a combination of linked-list and computed addressing, common in most computing processes, are called *plexes.*

Related to linked addressing is the concept of the *data graph,* whereby a block or data object is treated as a *node* or *vertex* and the address pointers called *edges, sides,* or *arcs* form the connections between nodes (Fig. 7-1). Unless each arc is represented by two pointers, so that the pointed-to block points in turn back to the preceding block in the list, the graph is *directed* and access can proceed only in one direction. The branching permitted by multiple pointers requires the differentiation of the two principal types of data graph, the *linear list* and the *tree.*

Lists, which can vary freely in length, vary as well in the ease with which blocks can be added or deleted. This flexibility is related to the pointers *within* the list as well as the available pointers provided *to* the list (Fig. 7-2). A *stack* is a list than can be accessed, with blocks added or deleted, at only a single end. A single address for the stack label points to the beginning of the list. A block is added by setting this pointer

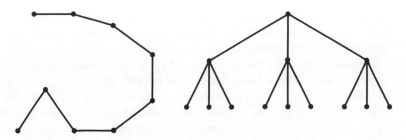

Figure 7-1 Data graphs showing linear list (left) and tree (right). Dots represent data objects linked by address pointers.

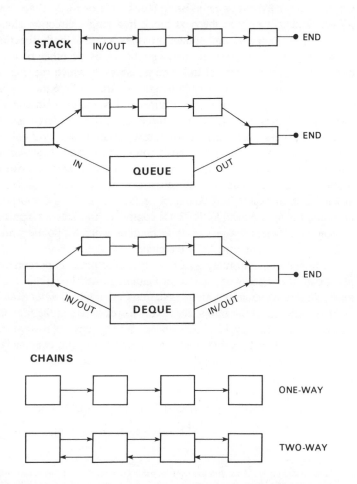

Figure 7-2 Linear list structures: stacks, queues, deques, one-way and two-way chains.

designating the stack to the address of the new block and by linking the pointer of the new block to the address of the previous first block. A block is removed by pointing the stack address to the second block. This type of list is similar to a stack of dinner plates raised and lowered in a cafeteria by a spring device so that only the top plate is visible. A stack is also called a LIFO list, for last in, first out.

More flexible access is provided by the queue, deque, and chain (Fig. 7-2). A *queue* is a FIFO list, for first in, first out, and is similar in organization to a waiting line at a theater box office. Blocks are stored at one end of the queue and retrieved at the other end; hence the queue must have addresses for both its beginning and end. A *deque,* analogous to a deck of cards in the hands of a less than scrupulous gambler, may have blocks added or deleted at either end. More versatile still is the *chain,* a plex with a linear ordering between blocks. Additions and deletions can occur anywhere by connecting in a new block or bypassing an existing block. In a *one-way* chain, each block has a single pointer linking it only to the next block in a single direction along the chain, whereas in a *two-way* chain a second pointer provides access in both directions. Although interior blocks may be inserted or deleted without removing other blocks from the chain, the chain often must be scanned to locate positions at which these changes are appropriate.

The advantages of a variable-length list structure are most appropriate in interactive digitization and editing of linear features and boundaries. In conventional programming these data objects are usually structured as vectors, ordered and fixed in length, with successive points stored in successive rows of an N row by 2 column array. A vector of length M $(M < N)$ occupies the first M rows of this fixed-length array, and if an additional point is to be added, all following points must be shifted downward to make room. Similarly, if a point is deleted, all subsequent points must be moved upward. The addition and deletion of points in a boundary or feature line is a cumbersome task if coded for the computer by a typical FORTRAN compiler. Interactive graphics systems rely instead upon more efficient list-processor techniques whereby pointers are merely reconnected to include or exclude a point in the chain.[1]

The hierarchical ordering of tree-structured data is appropriate for cartographic applications where areas or lines form natural, nested hierarchies. Clearly, the nesting of most administrative units follows a simple tree in which a *path* can be traced from a *root* at the national level through successively lower orders at the regional, state, county, and minor civil division levels. If pointers are directed upward toward the root, data for minor civil divisions can be aggregated into county totals and eventually a national summary can be produced. If downward pointers are provided, data for a particular small area can easily be retrieved by specifying a path such as

> U.S. > Northeast > Pennsylvania > Lackawanna County

> Scranton > Tract 12 > Block 18

---

[1]See Wolfgang K. Giloi, *Interactive Computer Graphics: Data Structures, Algorithms, Languages* (Englewood Cliffs, N.J.: Prentice-Hall, Inc., 1978), pp. 61–76.

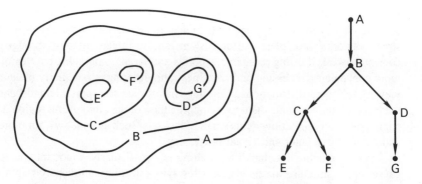

Figure 7-3 Nesting of contour lines (left) can be represented by a data tree (right). Arrows point upslope and represent two-way links.

A tree data structure can also represent the nesting of contour lines on a continuous topographic surface (Fig. 7-3). Trees as well as lists enjoy the flexibility of editing by redirecting address pointers between blocks.

Cartographic Data Manipulation

The value of a particular data structure must be judged according to the ways in which the information in the data base is likely to be manipulated. Like other data, geographic data are subject to the eight basic operations recognized by Page and Wilson:[2]

1. Accessing
2. Inserting
3. Deleting
4. Searching
5. Sorting
6. Copying
7. Combining
8. Separating

Accessing is basic to simple graphic displays, statistical analyses, and more complex cartographic and analytical procedures. In a cartographic context, inserting and deleting refer largely to editing, not only to correct mistakes but also to respond to a changing

---

[2]E. S. Page and L. B. Wilson, *Information Representation and Manipulation in a Computer*, 2nd ed. (Cambridge: Cambridge University Press, 1978), pp. 51–52.

landscape. Search is particularly important in many cartographic operations: to find sites suitable for a proposed development, to determine whether the file representing an urban street network is complete and consistent, and to thread contours through an elevation control network. Sorting reorganizes points and larger geographic entities in memory for more convenient processing and identifies rank orderings particularly meaningful to map viewers for some distributions. Copying makes data transportable and hence more valuable; combining permits the upward aggregation of data into useful regional and national summaries; and separating yields smaller data sets, such as windows, for more economical and detailed computer and visual processing.

For geographic data handling, editing is a particularly important consideration. Because a data structure that is efficient for symbolization and display might be unwieldy for correcting and updating, overall efficiency might mandate a duplication, with separate and differently organized files for different purposes. This duplication is promoted by easily segregated editing and analysis functions, which might well employ separate but communicating computers. The added cost of extra external memory devices frequently can be justified by lower costs for the central processor(s) and internal memory, a more rapid response of the system to user requests, and the accommodation of a larger number of users.

An Example

To illustrate the different demands of editing and display, consider a simple data structure intended for plotting choropleth maps (Fig. 7-4). Given the assignment of areal units to a small number of categories, the shading algorithm discussed in Chapter 5 generates the appropriate shaded area symbols as list D of (x, y, p) coordinates, with up-down code p controlling the plotting pen. For each areal unit the (x, y) coordinates of the polygon are copied into vector C, the initial data for the shading algorithm. All polygon points are stored sequentially in vector B. Each polygon is represented by a row in vector A,

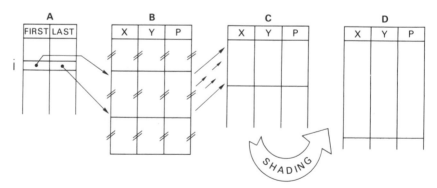

Figure 7-4 Shading coordinates in vector D are produced from polygon vector C, filled with coordinates from vector B as specified by pointers in vector A for area i.

with these rows ordered according to the alphabetical sequence of area names. In vector A an entry in the first column points to the row in vector B containing the first coordinate pair for the area in question, and an entry in the second column of vector A points to the row containing the last coordinate pair. Thus polygon i can be loaded into vector C for the shading algorithm by copying row $A(i, 1)$ of vector B into row 1 of C, and filling the rows of C sequentially with values from successive rows of B, through row $A(i, 2)$.

This organization of the data is inefficient for editing, particularly if the array organization common to FORTRAN is used for vectors A and B. A new areal unit might be added, for example, because a city has annexed a new subdivision that is to become a new census tract, or because an existing tract with a substantial influx of residents must be partitioned into two tracts. If the newly annexed subdivision's address pointers can be added at the current end of vector A and its coordinates appended to the end of vector B, the addition is relatively simple. If the insertion of these pointers must occur closer to the beginning of vector A, for instance, to accommodate an alphabetical sequence or an existing numerical ordering, subsequent rows must be moved downward in the vector before inserting the new pointers. Similarly, deletion of an area might require the shift upward of remaining rows in vectors A and B for computational neatness or to conserve storage.[3] The more obvious difficulties arise, of course, when individual points must be added or deleted; rows must be readjusted in vector B and address pointers reset in vector A.

Even the use of a chain-linked list would not obviate another difficulty in the editing of these data, the need to avoid overlapping area symbols or gaps along internal boundaries. If each polygon were digitized separately as a closed polygon, the overlap or gaps likely between adjoining, duplicate representations of common boundaries would be removed only by the most tedious human intervention or a complex algorithm. Ideally, each internal boundary should be represented only once to assure consistency in symbolization and to avoid useless redundancy.

Recognition of individual common boundary segments as basic data objects increases the complexity as well as the flexibility of the data structure (Fig. 7-5). Boundary segments might each be represented by a one-way chain so that points can be added or deleted with ease. A single chain CP can be used for all boundary-coordinate pairs, with another chain S of segment blocks containing, for each segment, pointers to the starting and ending points in coordinate chain CP. A third chain AS provides a clockwise sequence of boundary segments around each polygon as a series of successive blocks with reference pointers to the appropriate blocks in the segment chain S. Each run of blocks representing an areal unit in chain AS is identified, in turn, in area chain A by pointers to the first and last blocks of the run.

A chained list structure is particularly important for chains CP and AS, in which most errors are likely to occur. Points can be added to and deleted from the middle

---

[3]The values in these rows could be replaced with zeros, but in this case, every row retrieved from the vector would need to be inspected to determine whether it were an active row.

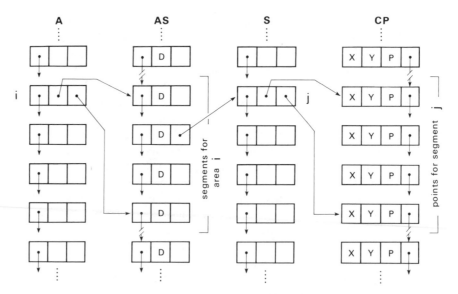

Figure 7-5 Coordinates for area polygon *i* are retrieved by following pointers in chain *A* to beginning and end of a run of blocks in chain *AS* representing segments forming perimeter of area *i*. Segment *j* in chain *S* has direction *D* for polygon *i* and is represented by a run of points in chain *CP*.

sections of segments solely by restructuring chain *CP*. Areas can be reconfigured by altering the linkages in chain *AS* to redefine the circuit of links around the perimeter of an areal unit. Similarly, linear features, for example, railroads and electrical transmission lines, can be represented and edited in chain *AS* as a series of segments.

Although providing the flexibility appropriate for editing, this chain data structure wastes computing effort and is awkward for generating and displaying choropleth shading patterns. Area-symbol algorithms operate efficiently on area polygons represented by vectors of points. For each area in list *A*, a vector similar to array *C* in Figure 7-4 must be developed by accumulating point coordinates for all segments defining the area in chain *AS*. Thus reference pointers must be followed from list *A* to chain *AS* to list *S* to chain *CP*, with pointers to successive blocks used in chains *AS* and *CP* to access all segments in each polygon and all points in each segment. This process is indeed cumbersome to repeat every time a choropleth map is generated, particularly when many maps are produced simply to experiment with the visual effects of different sets of class intervals.

A more efficient approach recognizes that the number of different shading patterns needed in traditional choropleth mapping is relatively small, perhaps, 10 to 20 at most. In practice, far fewer shadings are usually sufficient. In any event, why not compute in advance the point coordinates for each shading pattern in this repertoire of area symbols

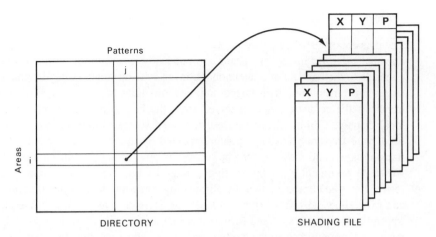

Figure 7-6 Pointer in shading directory locates shading pattern coordinates for pattern *j* of area *i*.

and provide the display portion of the system with a series of shadings from which a selection is made and plotted? The shading-pattern data thus can be organized sequentially, without the relatively complicated and time-consuming pursuit of address pointers between successive links in a chain. A *directory* of shading patterns is needed to provide for each area the address of the beginning of the shading coordinates for each pattern (Fig. 7-6).[4] Redundant calculations are eliminated, response time is reduced, the central processor is more available for other tasks, and the shading-symbol data can be forwarded to the software libraries of other display systems and enjoy a comparatively widespread use. To be sure, more external memory is needed than for the edit-efficient data structure in Figure 7-5, but the increased computational efficiency fully warrants providing this added storage.

Other efficiencies are possible by exchanging redundancy in calculations for redundancy in memory. If continuous-tone area symbols, as described in Chapter 5, are to be used for unclassed choropleth mapping, the full area polygons should be stored sequentially for ready access with a directory similar to that for the precomputed shading patterns. These polygon boundaries might also be made available for the plotting of boundaries, but as noted earlier, the storage of the individual segments comprising internal boundaries is more efficient because each common boundary need only be plotted once.

---

[4]The end of the series of coordinates for each shading pattern must be indicated either by a special, identifiable end code following the last meaningful coordinate in the shading list or by a second value in the pattern directory to indicate the number of points for the shading pattern.

Networking and Information Display

The concept of distributed data bases is an outgrowth of remote time-sharing computing.[5] Terminals can be linked to distant computers by a *modem* or an *acoustic coupler* that connects to the handset of a telephone, converts the terminal's digital signal to beeps and chirps for transmission over telecommunications lines to a distant central processor, and translates the acoustic pulses received from the host computer into digital displays for local display. Communications lines dedicated solely to data transmission can also be used, wire transmissions can be supplemented by microwave and other electromagnetic transmissions, and messages can be redirected by orbiting satellites to distant parts of the world. Government agencies sharing a central computer can also share the data bases residing on that computer, thereby eliminating some duplication in data-collection efforts and expanding the information available for decision making in any one agency. Data-transmission networks extend further this facility for data sharing by allowing one computer to communicate with another computer, and enabling one data-base system to receive information from another part of a *distributed-data-base* system.

An interesting development of the distributed data base concept is DIDS, the Decision Information Display System developed by NASA for the White House and maintained in the Department of Commerce.[6] A wide variety of federal agencies provide data for and have access to DIDS, and access can be provided to state legislatures and planning bodies, universities, and other nonfederal users.[7] Choropleth maps and statistical diagrams are displayed on a large-screen, color raster CRT unit with a "progressive zoom" feature for examining smaller areas in greater detail at large scales, and for regions, states, metropolitan areas, congressional districts, and individual cities, as well as for the nation as a whole at the state- and county-unit levels.[8] Video projectors can also be used to project maps and graphs onto a screen for viewing by many people in a large room.

Transmission of information through a telecommunications network can be uncomfortably slow to the user of an interactive system if a live video copy of a detailed map is to be transmitted for a high-resolution CRT display. Adding substantially to the response time would be the delay inherent in classing and symbolizing the geographic distribution on a large, centrally located, time-sharing system. The response time would be degraded further by a national crisis that would greatly increase the number of requests to the host system.

A far more efficient approach is planned for DIDS: sending out over the network to the requesting terminal only the vector of data values for the requested geographic

---

[5]See Robert J. Thierauf, *Distributed Processing Systems* (Englewood Cliffs, N.J.: Prentice-Hall, Inc., 1978), and Starr Roxanne Hiltz and Murray Turoff, *The Network Nation: Human Communication Via Computer* (Reading, Mass.: Addison-Wesley Publishing Co., 1978).

[6]See, for example, J. Dalton and others, "Interactive Color Map Displays of Domestic Information," *Computer Graphics,* 13, no. 2 (August 1979), 226–33.

[7]At the time of publication, DIDS is still considered an experimental program and is not yet widely available.

[8]Environmental data, for polygons other than the common statistical areas mentioned, might also be displayed by DIDS.

distribution rather than the symbol geometry for the map. Each terminal would be a "smart terminal," with a powerful microprocessor for data classification and symbolization in accord with the user's specifications. A copy of the base map would be stored on a floppy disk, to be "colored in" by the terminal. This segregation of the geographic base data and the areal-unit attribute data is characteristic of many geographic information systems because these two different types of file can be organized, captured, edited, and maintained largely separate from one another.

Worth emphasizing in the context of geographic data display networks is the superiority of raster data for fast and efficient graphic displays. Although raster information is sometimes discussed as if the data had to be structured in a grid format, with rectangular cells organized along rows and columns, a base map overlay for generating CRT displays is stored more efficiently according to the scan-line concept described in Chapter 2 for the CMAP line-printer mapping program.[9] Because the data are accessed sequentially row by row, but never up or down columns, not only is row-by-row storage appropriate, but within each row groups of contiguous "cells" representing the same areal unit can be assigned to variable-length scan-line segments. Scan-line coding thus reduces significantly the number of individual pieces of data that must be retrieved and processed in generating the raster display. The close correspondence in format between the scan-line data-base structure and the CRT screen assures computational efficiency.

TOPOLOGICAL CONSIDERATIONS

For some choropleth maps attention is most appropriately focused upon regional trends rather than upon providing a spatially ordered assemblage of symbol-coded data values. If the map author can assume that the map user will be more interested in observing that the higher values are largely in the Northeast and the Pacific Northwest and less interested in retrieving values for specific counties, the effectiveness of the map can be enhanced by not plotting internal boundaries between adjoining areal units assigned to the same category. A complete web of areal unit boundaries, while providing a useful geographic frame of reference for some map-reading tasks, can interfere with the perception and memory of regional trends, particularly if the internal boundaries are graphically dense, as on a page-size county-unit map of the entire United States.

Eliminating internal boundaries is simple for raster displays developed from scan-line data; these boundaries, which must be inserted deliberately between successive scan-line segments, simply are not added. Whereas the lengths of the scan-line segments provide geometric information about the map, the sequence of segments within the scan lines provides *topological* information, facts about the spatial units and their boundaries that are not affected by distortions unless the map is torn. Because raster files are top-

---

[9]The display of colored maps on raster display units is analogous to the overprinting of alphanumeric characters on the line printer by CMAP, as discussed in Chapter 2. The overprinted lines of a CMAP display are similar to the superimposed scan lines plotted by a color CRT unit for the three additive primary colors: blue, green, and red.

ologically ordered within each row, adjacencies among areal units can be recognized readily and internal boundaries inserted or omitted with ease.[10]

Vector polygons lack the convenient topological referencing inherent to raster data, and removal of internal boundaries is either computationally complex or impracticable without a data structure that provides information about adjacencies. Because areal units adjoin along boundary segments, the most logical and efficient means of adding topological referencing is to identify for each segment the areas lying on each side, the areas that *cobound* the segment. This information is conveniently added to memory blocks describing individual segments, as in list *S* in Figure 7-5. With this enhancement of the data base, internal boundaries can be plotted on a choropleth map only when the cobounding areas are assigned to different mapping categories. Contemplating the alternative of a costly and awkward algorithm for detecting and eliminating common boundaries should convince the most skeptical cartographer of the usefulness of topological referencing in a geographic data base.

DIME Features

Topological referencing has other, more far-reaching roles in computer-assisted cartography. A particularly useful refinement is the DIME (Dual Independent Map Encoding) concept developed by the U.S. Bureau of the Census to permit the semiautomatic editing of data bases describing the urban street network and statistical units such as blocks and census tracts. DIME files evolved from the Bureau's Address Coding Guides (ACGs), huge data bases used in the larger urban areas to assign block numbers to household addresses so that summary statistics can be aggregated for blocks and larger enumeration areas.

ACG-DIME data require that the urban area be represented by a network of points, line segments, and areas. Street segments between the intersections at opposite ends of a block form most of the links, but sections of administrative boundaries, railways, drainage lines, shorelines, and similar linear features are also used to bound census "blocks" for which aggregated data are to be tabulated (Fig. 7-7, left). Moreover, a curving street might be represented by several links to provide for a more realistic plotted map of straight-line segments.[11] All points at which two or more lines meet must be assigned unique node, or vertex, numbers. Similarly, segment (or edge) numbers and area (or polygon) numbers must also be unique.

The basic record in an ACG-DIME file describes each segment with the name of the street or feature, the ranges of addresses on both sides of the street, the node numbers for each end of the segment, and the polygon numbers of the cobounding areal units (Fig.

---

[10]Continuous-tone boundaries might be produced for classed choropleth maps to portray the relative lack of similarity between adjoining areal units in the same as well as in different categories, while permitting the principal area symbols to focus attention upon the regions defined by the classification.

[11]More graphically accurate representations of curvilinear features can be stored separately as chains for each link in the network.

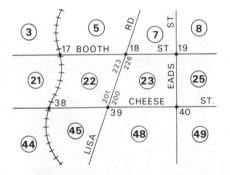

STREET NAME		Lisa
STREET TYPE		Road
ADDRESS RANGES:		
ODD	LOW	201
	HIGH	223
EVEN	LOW	200
	HIGH	226
LOW NODE		39
HIGH NODE		18
LEFT BLOCK		22
RIGHT BLOCK		23

Figure 7-7 Unique node and block numbers (left) are linked in DIME records for individual street segments (right) to provide a topological description of the network of segments, nodes, and areas.

7-7, right). These segment records thus permit the assignment of census counts for a particular household to the aggregate count for the appropriate block and tract. An address is first matched with the street name, then with the appropriate block on the basis of address range. Finally, the household count is added to the relevant block count for the odd- or even-numbered side of the street. Address data also can be matched with nodes and node coordinates, with linear interpolation used to estimate more accurate coordinates for mid-block addresses. ACG-DIME files thus can be used to prepare dot or point-symbol distribution maps for urban phenomena, as well as a variety of displays and tabulations for many different types of municipal, state, federal, and private-sector agencies concerned with public safety and the public welfare. In effect, the ACG-DIME file is a basic geographic cross-reference providing links between areal-unit, street-address, street-intersection, and geographic-coordinate data that enhance considerably the usefulness of data collected for cities.

Because "left" and "right" have significance only for a specific direction along a boundary, these directional references require a standard direction, as from the node at the low-address end of the block to the high-address node (Fig. 7-7). This relationship can be coded for a segment as the ordered quadruple

Node 1, Node 2 : Area A, Area B

to indicate that, in the direction from Node 1 to Node 2, Area A lies to the right and Area B lies to the left (Fig. 7-8, left). This relationship is symmetric in the sense that if the nodes are exchanged, the areas must also be exchanged to satisfy the structure among the elements specified by the quadruple. Thus the expression

Node 2, Node 1 : Area B, Area A

with exchanges for both sides of the colon, is still valid because, in the direction from Node 2 to Node 1, Area B lies to the right and Area A lies to the left (Fig. 7-8, right).

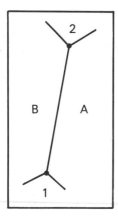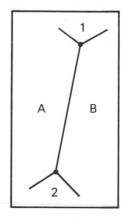

Figure 7-8 For direction from Node 1 to Node 2, Area A lies to the right and Area B lies to the left (left). Rotation of 180 degrees (right) shows symmetry of ordering, with exchanges of position between Areas A and B and between Nodes 1 and 2.

Computer algorithms use these directional codes to detect topological relationships among the nodes, segments, and areas in the DIME file.

The power of DIME encoding lies in the dualism of boundaries and coboundaries. In topological terms, these symmetric quadruples define the *ordered incidence relationships* among the *0-cells* (points) and (two-dimensional) *2-cells* (areas) that bound and cobound, respectively, the (one-dimensional) *1-cells* (segments) of a network (Fig. 7-9, left). Just as the two 0-cells bound a 1-cell directed from Node 1 to Node 2, two 2-cells can be said to cobound a 1-cell "directed" from Area 1 to Area 2 (Fig. 7-9, right). Thus the dual ordered incidence relationship for the coboundaries can be expressed as

Area A, Area B : Node 2, Node 1

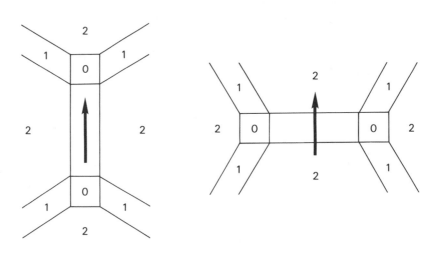

Figure 7-9 Ordered incidence relationships can be defined for 1-cell directed between two 0-cells (left) and its dual, a 1-cell directed between two 2-cells (right).

to indicate that, in the direction outward from Area A toward Area B, Node 2 lies to the right and Node 1 lies to the left. This orientation of nodes specifies a clockwise series of nodes, and by extension, segments around a polygon. The sequence of nodes around the perimeter of the polygon can be retrieved by detecting those segments outwardly cobounding an area polygon and linking together a clockwise chain of nodes.

Computer editing systems, such as the Bureau of the Census's ARITHMICON, can retrieve sequences of bounding 0-cells and cobounding 2-cells, inspect a DIME file for consistency, and identify errors that must be corrected.[12] The questions posed are as follows:

1. Is every 1-cell bounded by exactly two 0-cells?

2. Is every 1-cell cobounded by exactly two 2-cells?

3. Is every 2-cell bounded by a single closed, properly oriented sequence of 1-cells?

4. Is every 0-cell surrounded by a single closed chain of 2-cells cobounding the 1-cells linked to the particular 0-cells?

The second question is the dual of the first, and the fourth question is the dual of the third. By examining the quadruples representing the ordered incidence relationships among 0-, 1-, and 2-cells, a computer can determine if all answers are "yes," that a DIME file represents a topologically smooth surface, with no tears or discontinuities. Other topological tests are needed to ensure that the map is not a one-sided surface called a *Moebius strip* and that there are no "handles," as on a tea cup, caused by two nodes having the same node number.[13]

Land-use information systems based on vector data also can receive the benefits of DIME features for automated editing and analysis. The U.S. Geological Survey, for example, receives from its contractors raster-scanned data converted to vector mode for use with GIRAS, its Geographic Information Retrieval and Analysis System, and a flatbed plotter used to draw 1:250,000 and 1:100,000 land-use and land-cover maps. A DIME structure for these land-use polygons permits automatic editing and error detection and, in particular, the automatic chaining of unlabeled segments into polygons.[14] The 1-cells generally are curved lines called *arcs,* which must be represented by a list of point coordinates. For efficiency, GIRAS distinguishes between polygons and *islands,* 2-cells

---

[12]James P. Corbett, "Topological Principles in Cartography," in *Proceedings of the International Symposium on Computer-Assisted Cartography, Auto-Carto II, September 21–25, 1975* (Washington, D.C.: U.S. Bureau of the Census, and Falls Church, Va.: American Congress on Surveying and Mapping, 1977), pp. 61–65.

[13]Marvin S. White, Jr., "A Geometrical Model for Error Detection and Correction," in *Proceedings of the International Conference on Computer-Assisted Cartography, Auto-Carto III, January 16–20, 1978* (Falls Church, Va.: American Congress on Surveying and Mapping, 1979), pp. 439–56.

[14]Robin Fegeas, "Data Base Management System Approaches to Manipulation of Cartographic Data," in *Proceedings of the International Conference on Computer-Assisted Cartography, Auto-Carto III, January 16–20, 1978* (Falls Church, Va.: American Congress on Surveying and Mapping, 1979), pp. 457–71.

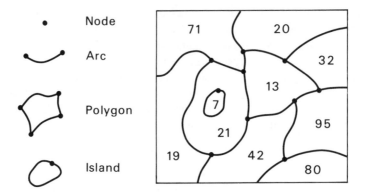

Figure 7-10 On digital, vector land-use maps, 2-cells are either polygons bounded by several arcs or islands bounded by a single arc.

completely surrounded by another 2-cell and bounded by a single arc closing upon itself at a single node (Fig. 7-10). Each polygon record includes the number, if any, of interior islands and, if the polygon is itself an interior island, identifies the single exterior polygon. Left-right designations of the nodes and areas bounding and cobounding the arcs complete the topological description. These modifications of the street-grid DIME structure provide an efficient accommodation of the geographic pattern of land-unit boundaries.

Topographic Data

Digitized topographic data can be organized in raster mode as an elevation grid, the most common type of digital terrain model, or in vector mode as contour chains. Because topologic relationships are implicit in their two-dimensional, row-and-column ordering, elevation grids are inherently efficient not only for comparisons with other data recorded for the same grid cells but also for automated spatial analyses involving small neighborhoods around grid cells, for example, the calculation of slope and aspect. In contrast, the only geographic ordering intrinsic to contour chains is the linear sequence of points along the contour line. Unless topologic attributes are added to the data structure, slope calculation and even the plotting of nonoverlapping labels is excessively cumbersome.

The addition of DIME features to vector representations of topographic and other continuous three-dimensional surfaces is simplified by the nesting and inherent tree structure of contour lines. Because contour lines do not intersect, the DIME concept of a 0-cell is not transferable.[15] Moreover, the concepts of "left" and "right" are less significant

---

[15]Contour lines could, of course, intersect in the neighborhood of a precipice, but this situation is rare. An analogy exists between the nodes of a network and the pits and peaks of a terrain surface and between the links in a network and geodesic paths. See Peter Haggett and Richard J. Chorley, *Network Analysis in Geography* (London: Edward Arnold, 1969), pp. 223–26.

than "upslope" and "downslope," which define the important directional relationships among contours. Furthermore, a closed contour can contain several other contour lines immediately upslope and several depression contours at the same elevation. If the contour line in question is a depression contour, interior contours can be depression contours at a lower elevation and standard contours at the same elevation. The multiple branching inherent in a tree data structure, together with specifications of the contours' elevations, permits the efficient processing of vector elevation data, as well as the automated detection of such inconsistencies as missing contours.

Contour chains, of course, lack the facility for geographic access by which a window specified with UTM, state plane, or spherical coordinates might easily be retrieved from an elevation grid. Display of windows covering comparatively small portions of the region represented in the data base can be expedited if each contour chain is identified by the coordinates of opposite corners of a bounding rectangle aligned with the coordinate axes. Efficiency can be further promoted by address pointers not only to the next contour upslope, but also the use of *index contours,* with pointers spanning, say, five or ten contour intervals so that the computer might more quickly determine which branches of the tree fall within the window and which contour chains must be clipped.

A similar hierarchical tree structure can expedite the retrieval and display of windows for a vector polygon file. Branching might be based upon an existing hierarchy of administrative units: states, counties, minor civil divisions, tracts, and blocks. With a land-information system for which this type of administrative hierarchy might be neither appropriate nor available for retrieving and displaying windows, a two-level hierarchy of area polygons conforming to existing land-use and land-cover boundaries, but approximating squares or slightly elongated quadrangles, might be useful (Fig. 7-11). In this fashion, area polygons can be clustered into large, somewhat irregular grid cells that

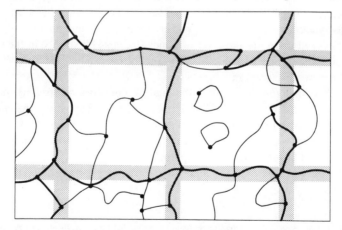

Figure 7-11 Irregular macro-polygons organized as a loose grid of higher-level areal units can provide land-use polygons with a weak but efficient tree structure, with easy retrievals for locations identified by grid coordinates.

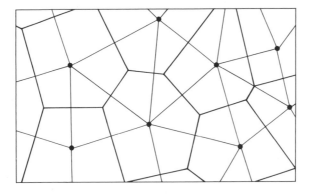

Figure 7-12 Network of triangles linking control-point nodes is the dual of the network of Thiessen, or Voronoi, polygons partitioning the area into proximal neighborhoods around the control points.

can readily be retrieved and matched with a specified window. This *loose grid* organization also promotes the analysis of neighborhoods around specific sites needed for environmental-impact assessment, because neighboring *macropolygons* can be referenced directly, as, for example, the cell in the next row to the south.

Tree structures can also provide a convenient, hierarchical organization for rectangular patches, or *unit frames,* that serve as basic building blocks for interactive displays. This concept of a *macrogrid* of vector data has been applied successfully by the Canadian Geographic Information System. A typical display includes a 32 by 32 subgrid of patches within the same window. A smaller subgrid would be used for a display at a larger scale, and a larger subgrid would be required for a display at a smaller scale.

Some types of isoline maps are based upon discrete control points: meteorogical stations, accurately surveyed control elevations, existing geological sample points such as outcrops and wells, isolated oceanographic sample points, and centers of area polygons. Isolines interpolated from these data to represent smooth, continuous three-dimensional surfaces can, of course, be stored and accessed as contour chains with a tree data structure. Subsequent analyses might, however, require the computation of additional or altered maps based upon the same set of control points. Flexibility in the geographic data base requires the storage of a network of triangles that can be used for positioning and threading isolines, as discussed in Chapter 5.

This triangular network is the dual of a network of proximal neighborhoods around control points, also called *Voronoi polygons* and *Thiessen polygons* (Fig. 7-12). These polygons partition the area into regions containing all points lying closer to the single control point within the region than to any other control point. The sides of these polygons, or their extensions, are also the perpendicular bisectors of links between control points, and these links serve as the sides of the triangles forming the dual network.[16] A simple

---

[16]For an algorithm to detect close pairs of nodes that might be linked to form a network of triangles, see J. L. Bentley and M. I. Shamos, "Divide-and-Conquer in Multidimensional Space," in *Proceedings of the 8th Annual ACM Symposium on the Theory of Computing* (New York: Association for Computing Machinery, 1976), pp. 220–30.

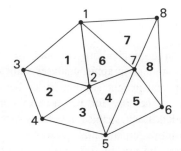

Δ Record	Associated Δs			Point Numbers		
	Δ1	Δ2	Δ3	1st	2nd	3rd
1	2	6		1	3	2
2	1	3		4	3	2
3	2	4		5	4	2
4	3	5	6	7	2	5
5	4	8		6	5	7
6	1	4	7	7	1	2
7	6	8		8	1	7
8	5	7		8	6	7

Figure 7-13 In the triangular network (left) control points 2 and 7 bound a link that is cobounded by triangles 4 and 6. In the topological table (right) triangle 4 is linked to associated triangles 3, 5, and 6, with opposite points 7, 2, and 5, respectively. Blank cells occur in the column of the third associated triangle for triangles on the edge of the network.

interpolation model can determine the intersections of isolines with these links, and the topologically coded network of triangles can be used to thread contours through intersections with the same elevation.

Although interpolation can be based solely upon the two control points bounding each link in the triangular network, a contoured surface more representative of the data and less responsive to minor, local trends would require a sample of at least four control points, the two points bounding the link and the opposite vertexes of the two triangles cobounding the link (Fig. 7-13). An intersection might then be found between a horizontal, parallel line above the link at the elevation designated for the contour and a quadratic surface fitted to these four data points according to least-squares methods. The contour line would then be threaded through the planimetric projection of this intersection.

A DIME-like file for the triangular network permits the identification of the opposite vertexes of the triangles cobounding the link.[17] Each triangle can be represented by a record containing the record numbers for the *associated triangles* sharing common boundaries with the triangle in question (Fig. 7-13, right). With these address pointers, a link between vertexes is represented by the pair of records for two adjoining triangles, that is, the two records containing the same pair of vertex numbers. The two vertexes directly opposite the link can then be retrieved as the remaining vertexes for each of these triangles.

This topological referencing of adjoining triangles also permits the threading of contour lines through the network. A contour that enters a triangle must exit through one of the two remaining sides, through the side with one node above and the other node below the contour elevation. The data record linking nodes and triangle numbers then identifies the next triangle to be entered, and the process continues until the contour

[17]M. H. Elfick, "Contouring by Use of a Triangular Mesh," *Cartographic Journal*, 16, no. 1 (June 1979), 24–29.

closes upon itself or encounters the edge of the map. As with the U.S. Census Bureau's ACG-DIME files, this digital representation of surface form also can be used during editing to detect topological inconsistencies.

These demonstrations of the usefulness of topological ordering and cross-referencing address pointers for cartographic retrievals and geographic analyses indicate that the design of geographic data bases requires special insight and expertise. Because spatial data handling often mandates redundancy in memory for the sake of computational efficiency, ample data storage must be provided. Because this redundancy extends beyond considerations of software and memory to more fundamental questions of machine architecture and computer operations, a geographic information system must be designed as a holistic system, with an equally thorough consideration of the system's uses, products, management, and personnel.

Cartography has a vested interest in the further development and implementation of data-base models. Data-base management systems (DBMSs) with straightforward, user-oriented languages have greatly simplified information storage and retrieval for many businesses. Yet data-base development commonly tends to address the comparatively simple, aspatial needs of inventory control and personnel administration, rather than the mixed locational and thematic attributes of geographic data. Particularly promising for geographic applications is the *relational data-base model,* based upon the theory of relational algebra. This concept is more abstract than those of existing data-base management systems and is not tied to particular programming systems. Theoretical studies indicate that with a relational data base attributes and relationships might be retrieved and manipulated in a comparatively simple fashion.[18] Much of the future success of geographic information systems will depend upon the spatial capabilities included in the design of content-addressable relational data bases.

---

[18]David Kroenke, *DATABASE: A Professional's Primer* (Chicago: Science Research Associates, 1978), pp. 149–84.

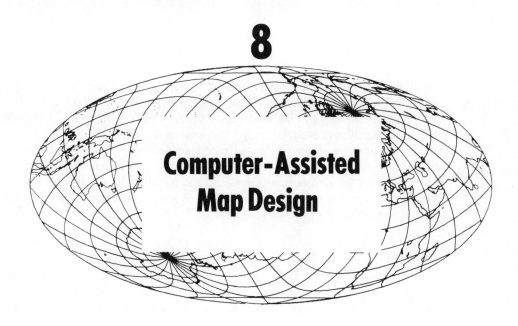

8

Computer-Assisted Map Design

What might be called the "digital transition" is shifting the labor in map making from people to computer-controlled machines. Despite these obvious trends toward increased automation, even the most ardent futurist will recognize that while mapping will make great strides as a "computer-assisted" process, "fully automated" mapping in the sense of, say, an automated brewing plant producing 500,000 liters of beer daily with only 10 production and supervisory employees is both unlikely and undesirable. The needs of the map user are too variable and unpredictable, and the consistency in flavor that can be an asset in the beer business can be downright stifling in a national mapping program, academic mapping, and private mapping firms. Fortunately, a computer-*assisted* cartography will help map makers to provide users with a responsive array of useful products as well as maintain high standards of accuracy.

Nowhere in any production process do the needs of the users influence the nature of the product more than in the design stage. The designer must originate ideas, plan all phases of production and distribution, and forecast the likely success of proposed products, both individually and in relation to the other activities of the firm. Design, after all, is a mental trial-and-error exercise intended to encourage innovation yet forestall a more costly and often embarrassing experimentation at the finished-product stage. This chapter surveys the potential role of the computer in planning and evaluating a map's graphic appearance, information content, and production program. If trial designs generated with a computer can be evaluated rapidly with either the designer's eyes or objective, calculated indexes of effectiveness, graphic failures and expensive errors can be avoided to the mutual benefit of the mapping agency and the user public.

GEOGRAPHIC CONTENT

Perhaps nowhere is thoughtful cartographic design more crucial than in the selection of a map's content: the theme or related themes portrayed, the region covered, and the data units employed. A diligent map user with a strong need to comprehend a geographic distribution usually will muddle through the tedious decoding of a visually complex, poorly symbolized, or sloppily executed and reproduced map. Yet, if an inappropriate distribution is portrayed, an important province excluded, a relevant related factor omitted, or a meaningless set of data units employed, the map user must either remain poorly informed, guess and thus risk misinformation, or search elsewhere. Good map design begins with meaningful map content.

Meaningful Areal Aggregations

What base maps are meaningful to the map user? For some purposes the fine resolution of a national map at the county-unit level or a state map at the minor-civil-division level may be more helpful than a display employing a coarser geographic framework, but not always; if decisions must be made about states, counties, or congressional districts, these areal units are likely to be more relevant to the decision makers. Similarly, urban-area data might be aggregated and displayed more meaningfully by ward, police precinct, or ethnic neighborhood than by census tract. Given an appropriate data structure, a computer can readily provide the appropriate upwardly aggregated data.

Meaningfulness depends upon the purpose of the map and an existing frame of reference. For many cartographic purposes, particularly for urban areas, a meaningful map is not available, perhaps because there is no consensus on what areal boundaries between, say, neighborhoods, are inherently meaningful. Because these judgments should be made by the people likely to use the map, a user survey might be in order. A sample of likely map users, or the whole population in the case, say, of a city council or other legislative body, can be asked to delineate on an outline map showing census blocks their opinions about appropriate neighborhood boundaries. A computer algorithm then measures the similarity for each pair of census blocks based upon the tendencies of the "judges" to assign both areas to the same neighborhood.[1] Factor analysis, a statistical method used to find groups of related variables, then extracts clusters of similar places representing neighborhoods.[2] Data aggregated to these *consensus neighborhoods* are more likely to provide a representative portrayal of socioeconomic data than census tracts commonly constrained to populations of approximately 4,000 residents.

---

[1]Mark S. Monmonier, "Pre-Aggregation of Small Areal Units: A Method of Improving Communication in Statistical Mapping," *Proceedings of the American Congress on Surveying and Mapping,* 35th Annual Meeting, 1975, 260–69.

[2]See R. J. Johnston, *Multivariate Statistical Analysis in Geography* (New York: Longman, Inc., 1978), pp. 127–82.

Data Reduction

In social science research, factor analysis is often employed for data reduction, with clusters of related variables collapsed to composite indexes and given names such as Socioeconomic Status, Occupational Prestige, or Segregation. These terms are usually meaningful to social scientists, who recognize the interrelatedness of many geographic distributions and who can benefit from viewing a few maps of distinctly different abstract concepts, rather than many maps of frequently redundant but more precisely defined geographic variables. Although the author of an atlas might appreciate the parsimony possible with factor analysis, the general reader is likely to be somewhat puzzled by the seemingly vague factor labels. Moreover, assigning titles to factors is by no means straightforward, a drawback further complicated by the need to choose among several methods for extracting factors from the data, as well as the need to specify how many factors to use.[3] The general atlas user, befuddled by vague factor labels, is unlikely to be further enlightened by a brief attempt to explain factor extraction.

A more realistic method of data reduction is based upon the *key-variable* approach to controlling the number of similar-looking maps in an atlas. The algorithm begins by computing the statistical correlations for all pairs of variables in the set of data and selecting for display the two most dissimilar geographic distributions.[4] All variables with a similarity to either of these distributions above a specified threshold correlation, say, 0.9, are then discarded. The next variable selected is the distribution with the least similarity to any of the selected variables. Any of the remaining variables within the similarity threshold of the newly selected distribution are then eliminated from further consideration, and this cyclic process continued until all variables are selected or discarded. Map titles can be comparatively straightforward; other important variables can be displayed if appropriate, perhaps at a smaller, less-detailed scale; and certain particularly important variables can be forced at the outset into the set of selected distributions.

Atlas Layout

The selection of maps for an atlas can be constrained by the organization of the atlas into sections and, if more than one map appears on a page, by sheet layout. There should, in accord with the objectives of the atlas, be a "balance" between the various sections. Moreover, as in the case of the second edition of the *National Atlas of the United States,* for which the individual map sheets are sold separately, an "integrity" must be maintained for each sheet: a topic requiring several smaller maps cannot, for example, begin near

---

[3]Risa Palm and Douglas Caruso, "Factor Labelling in Factorial Ecology," *Annals of the Association of American Geographers,* 62, no. 1 (March 1972), 122–23.

[4]Mark S. Monmonier, "Simplifying Data Reduction for More Effective Communication in Thematic Mapping," *Proceedings of the American Congress on Surveying and Mapping Fall Convention,* 1974, 104–15.

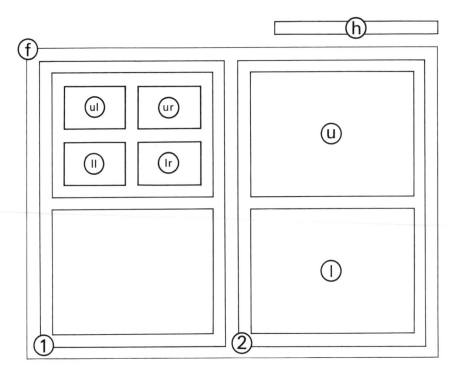

Figure 8-1 Example for nested hierarchy of layout codes for front face of map sheet.

the end of one sheet and terminate on the front of another sheet. Planning a detailed table of contents thus involves considerable trial-and-error experimentation with sheet layout.

This experimentation is assisted by the computer generation of simulated sheet layouts, maps of maps, in a sense, for preview individually at an interactive terminal or for hard-copy display with the line printer of sets of sheets.[5] Layout codes, stored in a data chain for easy insertion, deletion, transposition, and alteration by an interactive text editor, specify entries as sheetnames (s), header labels printed above the neat line (h), maps (m), text (t), or graphs (g).[6] A three-digit serial number identifies the sheet to which the entry belongs, and a scale code indicates whether a map, text, or graph is to occupy the entire page (7h), a quarter of the page (17), or a sixteenth of the page (34). For convenience these codes are based upon the denominators of the primary map scales covering these respective portions of the sheet, 1:7,500,000, 1:17,000,000, and 1:34,000,000. A hierarchy of location codes specifies either the front (f) or back (b) of

---

[5]Mark S. Monmonier, "Automated Techniques in Support of Planning for the *National Atlas*," *American Cartographer*, 8, no. 2 (October, 1981).

[6]A text editor is an interactive program for manipulating data by adding, deleting, or altering lines of text, parts of lines, expressions, and characters.

```
+-----------------------------++-----------------------------+
I text:                       II map:                        I
I Graph and Text              II Units Lacking Complete      I
I RE: Trends in Housing       II Plumbing Facilities as a    I
I     Quality                 II Percentage of All           I
I                             II Housing Units, 1980         I
I Rating:  A  B  C  D  Elim.  II (choro., counties)          I
I Comments:                   II                             I
I                             II Rating:  A  B  C  D  Elim.  I
I                             II Comments:                   I
I                             II                             I
I                             II                             I
I                             II                             I
I                             II                             I
I=============================II=============================I
I map:                        II map:                        I
I Units Lacking Complete      II Units Lacking Complete      I
I Plumbing Facilities as a    II Plumbing Facilities as a    I
I Percentage of All           II Percentage of All           I
I Housing Units, 1960         II Housing Units, 1940         I
I (choro., counties)          II (choro., counties)          I
I                             II                             I
I Rating:  A  B  C  D  Elim.  II Rating:  A  B  C  D  Elim.  I
I Comments:                   II Comments:                   I
I                             II                             I
I                             II                             I
I                             II                             I
I                             II                             I
+-----------------------------++-----------------------------+
```

Figure 8-2 Upper right quarter of a layout sheet showing header label for sheet face and four 1:34,000,000-equivalent sheet elements. (Courtesy U.S. Geological Survey.)

the sheet or the left (1) or right (2) half of the front face or the left (3) or right (4) half of the back face; the upper (u) or lower (l) portions of these half faces; and the upper right (ur) or left (ul) or lower right (lr) or left (ll) quarters of an upper or lower half-face (Fig. 8-1). These codes are used to position maps and other titles within a 120 line by 128 print-position array of alphanumeric characters representing the layout of an atlas sheet (Fig. 8-2). Layouts can be modified readily, and sheets requiring additional data or other selections from an abundance of possible maps can readily be identified.

MAP LAYOUT

Layout planning for the symbols and labels on a map is no less a problem than layout planning for an atlas. A variety of layout strategies exists, even for maps in a series that must conform to a standard format. Visual inspection and careful thought usually can yield a satisfactory layout, and the facility for repositioning map elements provided by an interactive graphics system is likely to encourage the experimentation and thought that are too easily stifled by the inertia of paper, pencil, and eraser in the hands of a map author. Yet for some complex layout tasks such as the positioning of names on an urban street map or the arrangement of areal units for a noncontiguous area cartogram, the tedious trial-and-error generation and evaluation of a large number of possibilities can better be left to a computer. A good cartographer has the capacity for good judgment, but the fatigue and boredom of label placement might well diminish this ability. Computers excel at boring, rote tasks and can be programmed to generate, evaluate, and modify layouts of map elements.

Computer-assisted placement of map components must consider two general types of element, movable and fixed. Movable elements include labels and certain point symbols that can occupy a variety of positions, yet must not overlap each other or fixed elements such as streets. A movable element might be separable, as are the individual words in certain labels; expandable in one direction, as are the letters in a word; and scalable, allowing enlargement or reduction without further directional distortion. Moreover, some labels and feature symbols can be mandatory, whereas others can be optional.

Elements can be bound to an area, a line, or a point. Curvilinear features might require corresponding curvilinear labels. Straight-line features usually require labels parallel to the features, but letter directions should be chosen to facilitate map reading, and positions both along and on an appropriate side of a feature can be chosen to minimize clutter. In some instances, the number of labels for a feature might also vary. In the spirit of cartographic license and in accord with the level of generalization for the map, symbols bound to a point might be shifted slightly away from the point, and labels representing an areal unit might be constrained not only to the interior of the polygon but also to a central portion of the area and to any dominant directional trend as well. Labels can also be aligned with parallels of latitude and restricted to a range of letter sizes.

A distinction should be made between constraints, which must be satisfied, and goals, which merely provide direction. Avoiding overlap is appropriately stated as a constraint, whereas the trade-offs that exist among many desirable design considerations are more suitably represented with relative weights in a composite *objective function*. This function measures the overall "goodness" of a solution and might be based upon the number of features that are labeled without overlap, considering as well the relative importance of each such feature and the extent to which the label meets letter-size, orientation, spacing, and other goals. A cartographic optimization model might thus acquire the precision and specificity of programming models in economics and business.[7]

Different solutions must be generated, tested against the constraints for acceptability, measured for optimality with the objective function, and discarded if a better solution already has been found. An initial feasible solution is required to initiate this series of iterations, and methods must be defined to generate new trial solutions and to recognize the point beyond which further trials are unlikely to yield an improved result. This *search strategy* should avoid *local optima,* solutions better than any further minor modifications might produce but still not the best possible solution, and seek instead the *global optimum,* a solution better than any local optimum.

---

[7]See Harvey M. Wagner, *Principles of Management Science with Applications to Executive Decisions,* 2nd ed. (Englewood Cliffs, N.J.: Prentice-Hall, Inc., 1975).

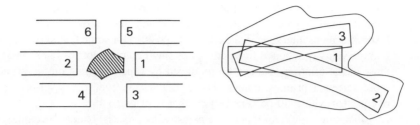

Figure 8-3 *Examples of priorities of name locations for a point symbol (left) and an areal unit (right).*

A computationally efficient approach to positioning labels and other movable map symbols would reduce the mind-boggling complexity of the problem to the manageable task of selecting among a limited number of label sites, each of which has a priority rating (Fig. 8-3).[8] Further economies would result from a hierarchical strategy: recognizing different levels of importance in the features to be labeled, obtaining an optimal solution first for major elements with only the most general consideration at this stage of where minor elements might be positioned, and fixing these more important labels before attempting to optimize the positions of less important labels. The higher-level labels would then be readjusted only if a satisfactory solution were not possible for the lower-level labels. Because a readjustment in one part of the map is unlikely to require alterations in more distant sections, a piecemeal approach is efficient and reasonable, particularly if the pieces are major areal units bounded by lines unlikely to be crossed by labels. Proceeding in this steady, step-by-step fashion, a properly guided computer can rapidly and consistently label a map with the care the conscientious cartographic draftsperson is unable or cannot afford to give the map.

PLANNING COLOR MAPS

Reproduction of a multicolor map requires careful planning and the coordination of a large number of *reproduction separates* or "flaps," sheets of scribecoat, peelcoat, photographic film, or transparent sheets holding labels. A complex production flow chart is usually needed to specify the interrelationships of these separates and the sequence of steps leading to the preparation of printing plates.[9] A different reproduction separate normally is prepared for each type of feature, even if two kinds of feature are to be printed in the same color.[10] A final negative, used to make a press plate, is prepared for each

---

[8]Pinhas Yoeli, "The Logic of Automated Map Lettering," *Cartographic Journal,* 9, no. 2 (December 1972), 99–108.

[9]See J. S. Keates, *Cartographic Design and Production* (New York: John Wiley & Sons, Inc., 1973), pp. 212–24.

[10]A separate for, say, principal roads might be used on another map that will include the separate with boundaries but omit those with railroads, buildings, contours, and the like.

ink color, and the composite positive used to make a final negative must collect through multiple photographic exposures all symbols and labels requiring the particular ink. Fine-textured screens intervene in the contact exposures for some features to reduce solid colors to more visually subdued fine patterns of small colored dots or lines, and hues other than those of the printing inks can be produced by superimposing on the printed map two or more screened primary colors. For example, the eye sees a 20-percent process blue overprinted with a 50-percent yellow as a light green.[11]

Planning a multicolor map is a painstaking process. Unless specifications are precise, labels that should be right-reading might be wrong-reading, light colors might be dark and dark colors light, and features that should not overlap might be superimposed. Moreover, the number of steps and flaps might be unnecessarily large, inflating the costs of labor and materials, and poorly chosen colors might readily be marred by minor misregistrations or overinking during printing, thus jeopardizing the visual effectiveness of the map.

If applied to map making, methods developed for computer-assisted design and industrial production planning can be used to prepare an efficient series of integrated reproduction steps from suitably coded specifications for map content and appearance.[12] Potential difficulties from misregistration of colors can be identified and suggestions posed for a more graphically stable design (Fig. 8-4). Production costs estimated by the computer for visually different designs can guide the writing of final specifications. An interactive design-assistance program can enhance considerably the cartographer's ability to cope mentally with a variety of minute but important decisions and to respond readily to the computer's suggestions for improvements.

The success of a map design is influenced by the geometry and topology of the base data. Although a design evaluation can be based in part upon the stability of finely detailed

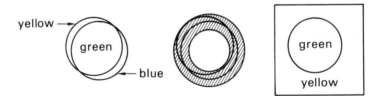

Figure 8-4 A green circle produced by overprinting blue and yellow screened tints on a white background might have diametrically opposite yellow and blue edges (left) as a result of misregistration during printing. This problem may be avoided by overprinting a black casing around the circumference of the circle (center) or printing a blue circle over a yellow background (right).

---

[11]The process inks are black and the subtractive primary colors: magenta, cyan, and yellow. Magenta is often called process red, and cyan is called process blue. Tint screens can mix these colors to duplicate most hues with a full range of brightness.

[12]For a discussion of computer-assisted design, see Warren J. Luzadder, *Innovative Design with an Introduction to Design Graphics* (Englewood Cliffs, N.J.: Prentice-Hall, Inc., 1975).

features under a small amount of misregistration of colors, it is more important to maintain sufficient visual contrast and a pleasing visual balance. Small amounts of bright purple, for example, can provide a useful contrast for certain linear features on a white, yellow, or orange background, yet large amounts of garish colors, particularly when juxtaposed with clashing colors, might be visually harsh and distract or repel the map reader. Objective functions and constraints based upon theories of color coordination and the cartographer's stated need to emphasize certain features can be used to optimize the selection of colors, as well as minimize registration difficulties.[13] A cartographic data structure with DIME or similar topologic features would permit the computer to inspect abutting symbols for color contrast and to determine the size of contiguous patches of the same color. Even if this process is not fully automated in an optimization model, map designers can benefit greatly from interactive software for rapidly identifying visually clashing colors and measuring visual balance.

Map reproduction is by no means foolproof, and many pitfalls exist besides the misregistration of colors. Not the least of the problems confronting the map author is the stability of black-and-white or color area symbols during printing. Most area symbols are reproduced by printing patterns of dots or lines intended to cover with ink only a designated percentage of the symbolized area. Invariably, these dots or lines emerge from the press larger than intended because, for a variety of reasons, more ink than necessary is deposited on the paper. A cartographer typically attempts to anticipate this expansion of the inked area, called *edge growth*, by selecting colors only after examining samples printed by the intended printers on paper similar to that for the map. Carefully monitoring the press and periodically checking ink coverage with a densitometer can also promote quality control. Nonetheless, some shifts in the hue and brilliance of colors and in the darkness of graytones are inevitable, and high-quality printing is costly. For many projects, particularly where printing quality is likely to be variable, the designer must anticipate and adjust for graphic noise resulting from overinking.

If the range of edge growth likely to occur or permitted by the printing contract is known or can be estimated, a computer can simulate the probable effects of overinking on each area symbol for a range of edge growths and thereby provide a more realistic assessment of visual stability and effectiveness.[14] Simulation models can estimate the likely effects of edge growth by determining the area covered by ink after generating dot diameters or line widths and separations, adding the estimated amount of edge growth to these diameters or widths, and calculating the new inked area while adjusting for the possible "overlap" of the expanded symbol elements. The cartographer can use graphs and tables based on simulated results to select more graphically stable symbols, such as a coarser dot screen or line screens with thicker lines, or to avoid area symbols likely

---

[13]See, for example, Johannes Itten, *The Art of Color*, Faber Birren, ed., and Ernst Van Hagen, trans. (New York: Van Nostrand Reinhold Company, 1970), and R. M. Taylor, "Empirical Derivation of Map Colour Specifications," paper presented at 9th International Conference on Cartography, College Park, Md., July 26–August 2, 1978.

[14]Mark S. Monmonier, "Modelling the Effect of Reproduction Noise on Continuous-tone Area Symbols," *Cartographic Journal,* 16, no. 2 (December 1979), 86–96.

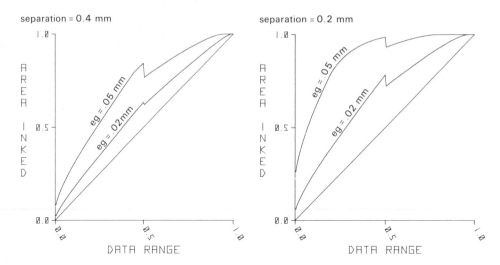

Figure 8-5 Simulations of edge growth incrementing dot radii by .02 mm and .05 mm for screens with rows of dots separated 0.4 mm (left) and 0.2 mm (right) illustrates the greater graphic stability of coarser screens. Most of the range of data values on the finer screen (right) is rendered black or nearly black by an edge growth of .05 mm.

to be rendered solid black by edge growth (Fig. 8-5). Designers of maps, like designers of buildings, engines, bridges, and pipelines, should consider all stresses, strains, and distortions likely to affect the materials with which a design becomes a reality.

CLASSIFICATION AND MAPPED PATTERN

Geographic patterns revealed on statistical maps may be simple or complex, easily described and memorable or jumbled and quickly forgotten. A mapped pattern might also be recognizably similar to a familiar distribution, and map symbols can portray data values for individual places with a high level of accuracy. Choropleth maps can consist of unclassed, continuous-tone symbols emphasizing values at places or classed symbols, limited in number and representing ranges of values for regions. The potential of classification in choropleth mapping to obscure meaningful details is too great to warrant a naive reliance on simplistic methods such as equal intervals or quantiles (equal ranks) for setting class breaks, and the conscientious map author must be aware of the extent to which a variety of choices would serve the objectives of the map.

Establishing class breaks that represent natural breaks in the data is a traditional goal in choropleth mapping. Although natural breaks might be blatantly obvious for some distributions, plotting the data values for the areal units as dots on a univariate axis, or *number line*, often fails to reveal a conveniently small number of distinct breaks. Nonetheless, if *m* classes are to be presented on the map, generally there is but one set of *m* − 1 class breaks that provides the most internally homogeneous mapping categories. Even though other considerations may be important to the map author, knowing the locations of these optimal breaks can be highly useful.

Given a suitable means of measuring the map designer's objectives, a computer can be programmed for the trial-and-error pursuit of an optimal solution to the class-interval problem. Classification error can be measured for each potential solution as the sum of the amounts by which every data value is misrepresented by the mean value $\bar{x}_j$ of the category *j* to which it is assigned. Dividing this result by the sum of the absolute deviations of each value from the overall mean $\bar{X}$ standardizes the total classification error to a range between 0 and 1, thereby providing a result that can be subtracted from 1 to provide a measure of relative accuracy, instead of an absolute measure of error. The resulting objective function is called the Tabular Accuracy Index (TAI) and is computed for *n* areal units as

$$\text{TAI} = 1 - \left[\frac{\sum\limits_{i=1}^{n} |x_{i(j)} - \bar{x}_j|}{\sum\limits_{i=1}^{n} |x_i - \bar{X}|} \right]$$

where data value $x_{i(j)}$ belongs to class *j* with mean $\bar{x}_j$.[15]

As with other optimization procedures, a set of constraints is needed to obviate unacceptable solutions. For a univariate classification, the most obvious constraints prohibit empty and overlapping categories. For certain applications the cartographer may want to specify one or more breaks inherently meaningful to some map readers. Zero, for example, distinguishes increases from decreases on a map of rates of change, and the national mean is useful as a standard against which to compare state or county values. The solution might be made more acceptable to some map users, as well as require fewer computations, if restricted to the limited number of round-number breaks. Although a rounded solution is likely to be less accurate than otherwise possible, this sacrifice of accuracy for simplicity is generally slight as long as the map author does not insist upon breaks with too few significant digits. In location-allocation modeling, of which optimal cartographic classification is a special case, a regional planner would dismiss a minimum

---

[15]See George F. Jenks and Fred C. Caspall, "Error on Choropleth Maps: Definition, Measurement, Reduction," *Annals of the Association of American Geographers,* 61, no. 2 (June 1971), 217–44. This method has been adapted to selecting representative symbols for legends of maps with proportional point symbols; see Michael W. Dobson, "Refining Legend Values of Proportional Circle Maps," *Canadian Cartographer,* 11, no. 1 (June 1974), 45–53.

transport site for a new landfill if that site were politically impracticable, as for example, an important historic site. Optimization should be a goal, not an obsession.

The look of a map is as important as internally homogeneous classes. Different sets of class breaks yield different visual patterns of choropleth symbols, some of which might be meaningful to the map user and provide as well a high, albeit suboptimal, level of tabular accuracy. In seeking an acceptable compromise between homogeneous categories and a revealing pattern, the map author might experiment with class breaks by viewing numerous maps, generated rapidly and inexpensively on a line printer or CRT. Yet, unless the designer can afford to devote considerable time to waiting for results, several shortcuts ought seriously to be considered. One vehicle for expediting this preview process is generalized base data of less computationally demanding area polygons having relatively few sides. Meaningful cartographic caricatures showing only cue features can yield statistical maps as informative for pattern previewing as more detailed displays.[16]

Additional shortcuts can accelerate pattern selection still further. A map author appreciating the potential value of a visually simple, comparatively unfragmented map pattern can invoke an optimization algorithm guided by a measure of pattern complexity. The fragmentation index, computed as

$$F = \frac{m - 1}{n - 1}$$

is a straightforward metric based on the number of areal units n and the number of map regions m. Each group of contiguous areal units assigned to the same category forms a map region. Because there will be at least one map region, subtracting one from both numerator and denominator yields an index ranging from 0, for a one-class map, to 1, for a highly fragmented map with no two adjoining areal units in the same category. This objective function would be minimized by an optimization procedure or used interactively to evaluate, without always displaying the pattern, the likely appearances of many trial maps (Fig. 8-6). A variety of indexes are available for measuring visual complexity, including indexes considering the relative sizes of map regions.[17]

A map author might also seek a map pattern similar visually to a given map of another distribution. Although this objective might be considered diabolical, perhaps should be viewed with skepticism, and surely should be used with discretion, cartographers occasionally have legitimate needs to create a strong visual impression of similarity in spatial pattern for two highly correlated distributions. An appropriate objective function for an optimization algorithm can be based upon the cross-classification array M, where M_{ij} is the number of areal units in class i of the distribution in question assigned to class

[16]Borden D. Dent, "A Note on the Importance of Shape in Cartogram Communication," *Journal of Geography,* 71, no. 7 (October 1972), 393–401.

[17]Mark S. Monmonier, "Measures of Pattern Complexity for Choroplethic Maps," *American Cartographer,* 1, no. 2 (October 1974), 159–69.

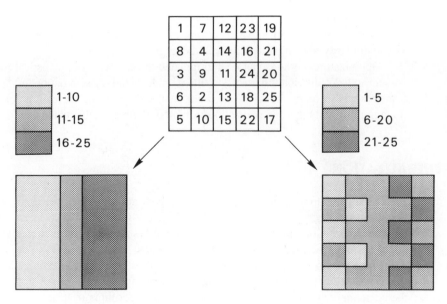

Figure 8-6 Uniform distribution of numbers can produce highly fragmented pattern with $F = .625$ (right) or comparatively simple pattern with $F = .083$ (left).

j on the given map of the referent distribution. A matching coefficient c_i can be computed for each of k classes as

$$c_i = 0.5 \frac{M_{ii} - Q_i}{M_{ii} + Q_i}$$

where Q_i, computed as

$$Q_i = \sum_{j=1}^{k} M_{ij} \qquad (i \neq j)$$

is the number of places assigned to class i for the variable under consideration but not placed in the corresponding class of the map for the referent distribution.[18] Overall visual similarity can then be measured by the weighted average S of these class-specific matching coefficients, calculated as

$$S = \frac{\displaystyle\sum_{i=1}^{k} a_i c_i}{\displaystyle\sum_{i=1}^{k} a_i}$$

---

[18]Mark S. Monmonier, "Class Intervals to Enhance the Visual Correlation of Choropleth Maps," *Canadian Cartographer*, 12, no. 2 (December 1975), 161–78.

where the weights a_i are the aggregate areas of mapping units assigned to class i on the new map. Composite objective functions can be used to base the selection in part on visual similarity and in part on internal homogeneity.

Optimization methods can be extrapolated to other types of statistical map, such as isoline and dot-distribution maps. With a widespread dissemination of software and firmware, these tools should be as much at the disposal of the cartographer as linear programming and other resource-allocation models are available to the economic planner.

GENERALIZATION

A suitable geographic frame of reference is important for any map. Even though the base data must usually be smoothed and filtered for portrayal at smaller scales, or for more rapid response with less flicker on an interactive system, this elimination and refinement ought not be haphazard and might even be pursued in its own right to adjust the detail of the base data to the map's scale of concepts. Indeed, base data generalization is perhaps the most intellectually challenging task for the cartographer, a proposition supported by the comparatively marginal success of computer algorithms in generalizing maps. This generalization gap in computer-assisted cartography stems not only from the computational complexity of the problem but also from only a vague understanding of the objectives and principles of map simplification. The intellectual components of this largely intuitive process must first be distilled from the more rote, easily programmable tasks before much progress will occur.

Although generalization can be extended to include symbolization and data classification, generalization of the base data for a reduction in scale is related largely to three processes: feature selection, geometric smoothing, and feature shifting. Geometric smoothing is needed to avoid the inelegant and distracting bumps and blemishes that would result from scaling downward tight curves and small wiggles. Elimination of less significant features does not always guarantee the avoidance of clutter, as when an important railway, highway, and river share the same passage through a mountainous region and must be pushed apart as well as smoothed. Feature shifting is also needed to preclude the overlapping of lines on opposite sides of the same feature such as a small island or narrow peninsula. A fully effective generalization algorithm must coordinate all three operations.

Feature Selection

In its most elementary form, computer-assisted feature selection depends upon feature codes that incorporate a ranking of features providing priorities so that a sufficient number of less important types of features can be suppressed to avoid cluttering the map plotted at a reduced scale. Coding streams according to branching characteristics, for example,

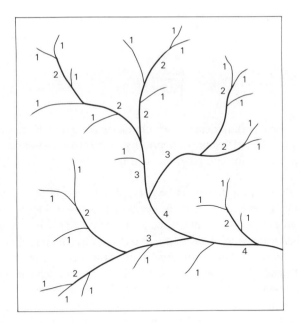

Figure 8-7 Drainage hierarchy based on branching characteristics (left) can be used to eliminate less significant stream segments for display at a smaller scale (right) in which all first order streams have been omitted.

provides for the exclusion of some probably minor drainage lines (Fig. 8-7). Highways can be assigned priorities based upon the traditional road hierarchy from Interstate routes, downward through U.S. routes, state routes, and county routes to town roads and private lanes, and railroads might be ranked as single-, double-, and multiple-track. More meaningful feature codes, of course, might consider stream discharge and traffic flow, yet this seemingly objective approach is not without pitfalls. For example, in a large and diverse region, a given stream discharge that would warrant removing a sluggish stream in a humid area might well eliminate a comparatively prominent fluvial landmark in a semiarid area. Variable thresholds are traditional in cartographic generalization, and have been applied to both point phenomena and linear features. As an example, Baltimore, Maryland, a thriving city about 80 km (50 miles) north of Washington, D.C., and 160 km (100 miles) south of Philadelphia, Pennsylvania, is commonly omitted from small-scale world maps and globes and enjoys the dubious distinction of being the largest unknown city in America.

A computer algorithm can implement the concept of the variable threshold more easily and more effectively than can the cartographer. The extent to which details can be retained might be specified with formulas similar to the uniform-density "law" derived from Töpfer and Pillewizer to relate the number n_f of features on a new map at scale M_f to be retained from a source map at scale M_a having n_a features.[19] Yet their formula

$$n_f = n_a \left(\frac{M_a}{M_f} \right)^{\frac{1}{2}}$$

does not directly address local feature density, which relates more directly to map clutter than does the aggregate number of features. A priority-based, trial-and-error procedure not unlike that described earlier for computer-assisted label placement would provide a more suitable solution.

Nevertheless, feature selection should not be pursued independent of feature shifting and geometric smoothing. Overlapping of point symbols and linear features, including the boundaries of some area symbols, might best be avoided by suppressing some features completely, plotting other features in place, and moving still other features slightly from their accurate locations. As is common with large-scale maps prepared from aerial photographs, map accuracy standards can be expressed as the maximum percentage, say, 90 percent, of features that might deviate from their accurate locations by more than a stated distance, such as 0.5 mm (0.02 in.) for maps at scales smaller than 1:20,000.[20] A map user can thus be provided with an appropriate uniformity in both accuracy and useful detail.

Line Smoothing

Lines selected for display, whether for linear features or as boundaries of symbolized areas such as parks and built-up areas, frequently must be smoothed to minimize distracting irregularities when the scale is reduced. Most computer-controlled attempts to smooth lines have been developed as vector-mode algorithms for eliminating points from data chains captured with a digitizer and intended for display with a pen plotter or vector CRT unit. These line simplification algorithms have been highly simplistic, for example, selecting every kth point or eliminating all points closer to a previously selected point than a set threshold distance. These approaches are useful in reducing substantially the number of points captured by a digitizer recording coordinates continuously, say, at intervals of half a second or smaller. Nonetheless, the cartographer must select with care the sampling interval k or distance threshold in order to avoid the arbitrary elimination of important parts of features. If the points are not spaced uniformly, even selecting every

---

[19]F. Töpfer and W. Pillewizer, "The Principles of Selection, A Means of Cartographic Generalisation," *Cartographic Journal*, 3, no. 1 (June 1966), 10–16.

[20]A statistical approach based on sampling theory might be more appropriate than the present method for specifying the accuracy of large-scale maps; see Morris M. Thompson and George H. Rosenfield, "On Map Accuracy Specifications," *Surveying and Mapping*, 31, no. 1 (March 1971), 57–64.

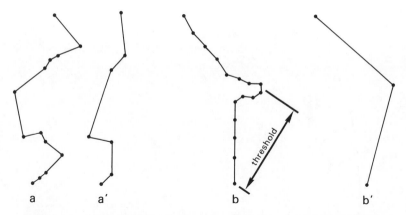

Figure 8-8 Simplistic generalization algorithms can simplify linear features by (a) eliminating every other point or (b) retaining only points a set threshold from a previously selected point.

a a′ b b′

other point might well distort the essential form of a line (Fig. 8-8, left). Furthermore, a set distance threshold large enough to remove redundant points along a relatively straight section of a line might not preserve properly a significant change in direction or a small but important part of the feature (Fig. 8-8, right).

A composite algorithm considering both proximity to other points and directional trends is more desirable than an approach based upon only a single criterion. Although direction and distances between adjacent points are not considered directly, an interesting algorithm advanced by Douglas and Peucker attempts to preserve directional trends and permits the cartographer to specify the extent to which a linear feature is simplified.[21] The two end points of the line are selected for display. Distances are then computed between the intervening points and a straight line connecting these end points (Fig. 8-9). If the greatest perpendicular distance from the straight line to any of these points is less than a threshold distance, all intervening points can be eliminated from the generalized line. If the threshold distance is exceeded, the point most distant from the straight line is selected for display and is used as an end point for the two resulting subdivisions of the original line. Each of these portions is then examined separately, with its own hypothetical straight line and recomputed perpendicular distances to intervening points. The process continues until no further points are available for elimination.

Selection of a suitable distance threshold can be based upon the width of the line at the reduced scale. Recognition of the finite thickness of a cartographic line as distinct from the infinitesimal width of a geometric line is an important consideration in map

---

[21]David H. Douglas and Thomas K. Peucker, "Algorithms for the Reduction of the Number of Points Required to Represent a Digitized Line or Its Caricature," *Canadian Cartographer*, 10, no. 2 (December 1973), 112–22.

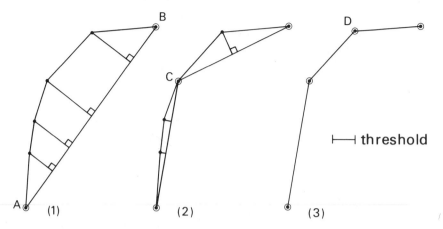

Figure 8-9 Successive selection of points *A, B, C,* and *D* by the Douglas-Peucker algorithm for line smoothing.

generalization, particularly if comparatively complex double-line and dashed-line symbols are to be used.[22]

The deviation of the generalized line from the path of the ungeneralized line provides another useful approach to controlling the amount of smoothing. George Jenks and his colleagues, for example, have measured generalization error as the area between the original and smoothed lines (Fig. 8-10, left). Other measures of correspondence can also measure the extent to which a simplification algorithm preserves directional trends, wiggliness, positional accuracy, and other characteristics of lines. Directional similarity is particularly important for lines that must be shifted apart to preclude overlap (Fig. 8-10, right).

The concept of generalization error is vaguely similar to the statistical notion of error used in least-squares curve fitting. These two concepts merge in automated line generalization because lines stored as vectors of discrete points must be displayed as smoothed curves through spline interpolation. Some points might be fixed to serve as anchors at which two curves approximating adjoining portions of the same feature meet, whereas other points might only be approximated by the smooth curves generated mathematically to provide a simplified representation of the original vector. In this case the degree of generalization might be measured by the average distance between these intermediate points and the closest approach of the smooth curve. Line thickness as well as the difference in detail between original and final scales might be considered by constraining the curvature of the fitted curved line segments so that, for example, the opposite sides of a small peninsula do not approach within a set distance. Each arc has a minimum radius of curvature, which might be constrained not to fall below a minimum radius consistent with the scale and line width of the map.

[22]Thomas K. Peucker, "A Theory of the Cartographic Line," *International Yearbook of Cartography*, 16 (1976), 134–43.

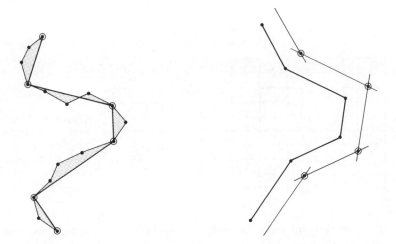

Figure 8-10 Shaded area indicates generalization error, or lack of locational similarity, between 16-point line and its 6-point generalization (left). Directional similarity is important for a line generated a constant distance to the right of an existing line (right).

Feature Shifting

Shifting features from their geometrically accurate positions can frequently prevent clutter that might be caused by a reduction in scale, and for many applications a more reasonable approach to generalization would first consider a minor relocation before deciding to eliminate a map feature. Line-simplification algorithms treating features as isolated entities unconstrained by other map elements are impractically shortsighted.

Because the spatial searches needed to avoid overlap are computationally complex in vector mode, the potential of a raster-mode approach to line generalization, used either alone or jointly with vector-mode calculations and data in a hybrid algorithm, merits consideration. The simplest raster-mode approach to generalization eliminates alternate rows and columns in a square grid while maintaining the connectivity of important lines. Shifts are needed to preserve connectivity when two or more lines a single cell wide touch. Different smaller scales can be attained by dropping only every kth row and column, and the range of possible scales is related inversely to cell size.

A set of rules to guide the shifting and elimination of cells representing features can, in a sense, serve as a grammar for map generalization. These rules might be stated as pairs of subgrids, 3 by 3 and larger. One element of the pair contains the initial configuration of cells and is matched to a corresponding portion of the grid map, for which the other part of the pair specifies the generalized configuration of cells (Fig. 8-11). A variety of different rules, including transformations useful for aggregating nearby but discontiguous areal units with similar land covers into larger, more cohesive generalized units, is possible. A hierarchy of operations might be appropriate, with matches

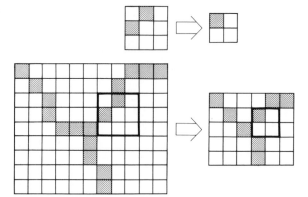

Figure 8-11 Transformation pairs similar to example (above) can be applied to generate smoothed features for reduced-scale map produced by eliminating alternate rows and columns (below).

attempted first with more commonly required rules. Substantial reductions in scale might be attained by several successive steps at which alternate rows and columns are removed. Overall efficiency could be improved by a machine architecture developed particularly for processing large arrays.[23]

Generalization and Intelligence

It is doubtful that the fundamental characteristics of most geographic features can be simplified in a completely satisfactory fashion merely by eliminating or shifting points or grid cells. The goal of map generalization is the preservation of the essential characteristics of map features, and for some features this essence is very much dependent upon map scale. The coastline of Maine, for example, is more distinguished by its drowned valleys and offshore islands than by its roughly northeast-southwest trend. For some map scales a truly reliable generalization algorithm would represent the Maine coast by eliminating some of these valleys and islands so that others could be exaggerated slightly. A point-by-point approach is hopelessly nearsighted because the algorithm must first subdivide the line into these fundamental units.

The automated recognition of essential feature characteristics is a major obstacle to fully automated cartographic generalization. The cartographer with the appropriate cognate skills in geomorphology and regional geography will understand the geometry of the Maine coast and, unless the scale is too small, will attempt to enhance some of these salient elements while suppressing others for the sake of clarity. It is doubtful that a mapping agency would ever seriously consider a proposal to endow a computer system with a similar capacity for interpreting the landscape if the principal benefit would only be a more rapidly produced, slightly more consistent map.

Perhaps nowhere in the interplay among map author, computer, and map is the term

---

[23]See, for example, W. R. Cyre and others, "WISPAC: a Parallel Array Computer for Large-scale System Simulation," *Simulation,* 29, no. 5 (November 1977), 156–72.

"computer-assisted" instead of "automated" more germane than in cartographic generalization. The most fully automated and still practicable approach would appear to require the intelligent guidance of a cartographer with an interactive CRT display and light pen. This strategy is well accepted in remote sensing for the multispectral classification of land-cover information. For map generalization, the map author might make certain primary decisions for a set of typical areas, and the computer would then replicate these decisions elsewhere throughout the region. Further advances in computer architecture for graphic applications, as well as a fuller understanding of the geometric properties of vector features and raster maps, will enable both map maker and map user to receive more completely the precision and cost savings of modern computer technology.

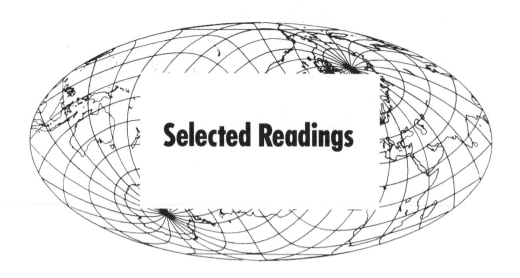

Selected Readings

1 An Introduction

ADAMS, T. A., H. M. MOUNSEY, and D. W. RHIND, "Topographic Maps from Computer Output on Microfilm," *Cartographic Journal*, 17, no. 1 (June 1980), 33–39.

BELL, S., and P. A. WOODSFORD, "Use of the HRD-1 Laser Display for Automated Cartography," *Cartographic Journal*, 14, no. 2 (December 1977), 128–34.

BOYLE, A. R., "Development in Equipment and Techniques," *Progress in Contemporary Cartography*, 1 (1980), 39–57.

———, "Automated Cartography," *World Cartography*, 15 (1979), 63–70.

———, "Geographic Information Systems in Hydrography and Oceanography," *Canadian Cartographer*, 11, no. 2 (December 1974), 125–41.

———, "The Requirements of an Interactive Display and Edit Facility for Cartography," *Canadian Cartographer*, 13, no. 1 (June 1976), 35–59.

BRASSEL, KURT, "A Survey of Cartographic Display Software," *International Yearbook of Cartography*, 17 (1977), 60–77.

BROOKS, FREDERICK P., JR., *The Mythical Man-month: Essays on Software Engineering*. Reading, Mass.: Addison-Wesley Publishing Company, Inc., 1975.

DAVIS, JOHN C., and MICHAEL J. McCULLAGH, eds., *Display and Analysis of Spatial Data*. New York: John Wiley & Sons, Inc., 1975.

DAVIS, SAMUEL, *Computer Data Displays*. Englewood Cliffs, N.J.: Prentice-Hall, Inc., 1969.

HERTZ, C. H., and T. ORHAUG, "The Ink Jet Plotter: A Computer Peripheral for Producing Hard Copy Color Imagery," *Computer Graphics and Image Processing*, 5, no. 1 (March 1976), 1–12.

KAZAN, B., "Materials Aspects of Display Devices," *Science*, 208, no. 4446 (23 May 1980), 927–36.

KEATES, J. S., *Cartographic Design and Production*, pp. 143–50, 182–94. New York: John Wiley & Sons, Inc., 1973.

MARBLE, DUANE F., and OTHERS, *Computer Software for Spatial Data Handling*, 3 volumes. Ottawa: International Geographical Union, Commission on Geographical Data Sensing and Processing, 1980.

MEYER, MORTON A., FREDERICK R. BROOME, and RICHARD H. SCHWEITZER, JR., "Color Statistical Mapping by the U.S. Bureau of the Census," *American Cartographer*, 2, no. 2 (October 1975), 100–117.

MOELLERING, HAROLD, "Strategies of Real Time Cartography," *Cartographic Journal*, 17, no. 1 (June 1980), 12–15.

MONTUORI, JOHN S., "Image Scanner Technology," *Photogrammetric Engineering and Remote Sensing*, 46, no. 1 (January 1980), 49–61.

MORRISON, JOEL L., "Computer Technology and Cartographic Change," *Progress in Contemporary Cartography*, 1 (1980), 5–23.

———, "Changing Philosophical-Technical Aspects of Thematic Cartography," *American Cartographer*, 1, no. 1 (April 1974), 5–14.

NEWMAN, WILLIAM M., and ROBERT F. SPROULL, *Principles of Interactive Computer Graphics*, 2nd ed. New York: McGraw-Hill Book Company, 1979.

PETRIE, G., "Hardware Aspects of Photogrammetry," *Photogrammetric Engineering and Remote Sensing*, 47, no. 3 (March 1981), 307–20.

PEUCKER, THOMAS K., *Computer Cartography*. Commission on College Geography Resource Paper No. 17. Washington, D.C.: Association of American Geographers, 1972.

PRENIS, JOHN, *Running Press Glossary of Computer Terms*. Philadelphia: Running Press, 1977.

RHIND, DAVID, "The Nature of Computer-Assisted Cartography," *Progress in Contemporary Cartography*, 1 (1980), 25–37.

ROBINSON, ARTHUR, RANDALL SALE, and JOEL MORRISON, *Elements of Cartography*, 4th ed., pp. 259–78. New York: John Wiley & Sons, Inc., 1978.

TEICHOLZ, ERIC, and JULIUS DORFMAN, *Computer Cartography: World-Wide Technology and Markets*. Newton, Mass.: International Technology Marketing, 1976.

THOMPSON, MORRIS M., *Maps for America: Cartographic Products of the U.S. Geological Survey*. Washington, D.C.: U.S. Government Printing Office, 1979.

WASTESSON, OLAF, BENGT RYSTEDT, and D. R. F. TAYLOR, eds., "Computer Cartography in Sweden," *Cartographia,* monograph no. 20, 1977.

2 Computers and Algorithms

BERGERON, R. DANIEL, PETER R. BONO, and JAMES D. FOLEY, "Graphics Programming Using the Core System," *Computing Surveys,* 10, no. 4 (December 1978), 389–443.

DAWSON, JOHN A., and DAVID J. UNWIN, *Computing for Geographers.* New York: Crane, Russak & Co., Inc., 1976.

GILOI, WOLFGANG K., *Interactive Computer Graphics: Data Structures, Algorithms, Languages.* Englewood Cliffs, N.J.: Prentice-Hall, Inc., 1978.

MacDOUGALL, E. BRUCE, *Computing for Spatial Problems.* London: Edward Arnold (Publishers), Ltd., 1976.

NAGY, GEORGE, and SHARED WAGLE, "Geographic Data Processing," *Computing Surveys,* 11, no. 2 (June 1979), 139–81.

RHIND, DAVID, "Computer-aided Cartography," *Transactions of the Institute of British Geographers,* new series, 2, no. 1 (1977), 71–97.

WEBSTER, MARTIN, "Buying a Microcomputer," *Journal of Geography in Higher Education,* 4, no. 1 (Spring 1980), 42–50.

WHITE, ROBERT M., "Disk-Storage Technology," *Scientific American,* 243, no. 2 (August 1980), 138–48.

3 Raster Symbols and Surface Mapping

BRAILE, LAWRENCE W., "Comparison of Four Random to Grid Methods," *Computers and Geosciences,* 4, no. 4 (1978), 341–49.

CERNY, JAMES W., "Use of the SYMAP Computer Mapping Program," *Journal of Geography,* 71, no. 3 (March 1972), 167–74.

EDWARDS, KATHLEEN, and R. M. BATSON, "Preparation and Presentation of Digital Maps in Raster Format," *American Cartographer,* 7, no. 1 (April 1980), 39–49.

LIEBENBERG, ELRI, "Symap: Its Uses and Abuses," *Cartographic Journal,* 13, no. 1 (June 1976), 26–36.

MORRISON, JOEL L., "Observed Statistical Trends in Various Interpolation Algorithms Useful for First Stage Interpolation," *Canadian Cartographer,* 11, no. 2 (December 1974), 142–59.

PEUCKER, THOMAS K., "The Impact of Different Mathematical Approaches to Contouring," *Cartographica,* 17, no. 2 (Summer 1980), 59–72.

PEUQUET, DONNA J., "Raster Processing: An Alternative Approach to Automated Cartographic Data Handling," *American Cartographer,* 6, no. 2 (October 1979), 129–39.

PROUZET, J., "Estimation of a Surface with Known Discontinuities for Automatic Contouring Purposes," *Mathematical Geology,* 12, no. 6 (December 1980), 559–75.

RHIND, D., "A Skeletal Overview of Spatial Interpolation Techniques," *Computer Applications,* 2, nos. 3–4 (1975), 293–309.

————, "Automated Contouring—an Empirical Evaluation of Some Differing Techniques," *Cartographic Journal*, 8, no. 2 (December 1971), 145–58.

4 Raster-Mode Measurement and Analysis

COLVOCORESSES, ALDEN P., "Evaluation of the Cartographic Applications of ERTS-1 Imagery," *American Cartographer*, 2, 1 (April 1975), 5–18.

FREEMAN, HERBERT, "Computer Processing of Line-Drawing Images," *Computing Surveys*, 6, no. 1 (March 1974), 57–97.

JENSEN, JOHN R., "Digital Land Cover Mapping Using Layered Classification Logic and Physical Composition Attributes," *American Cartographer*, 5, no. 2 (October 1978), 121–32.

LILLESAND, THOMAS M., and RALPH W. KIEFER, *Remote Sensing and Image Interpretation*. New York: John Wiley & Sons, Inc., 1979.

MONMONIER, MARK S., "Digitized Map Measurement and Correlation Applied to an Example in Crop Ecology," *Geographical Review*, 61, no. 1 (January 1971), 51–71.

————, JOHN L. PFALTZ, and AZRIEL ROSENFELD, "Surface Area from Contour Maps," *Photogrammetric Engineering*, 32, no. 5 (May 1966), 476–82.

MULLER, JEAN-CLAUDE, "Map Gridding and Cartographic Errors: A Recurrent Argument," *Canadian Cartographer*, 14, no. 2 (December 1977), 152–67.

NORDBECK, STIG, and BENGT RYSTEDT, *Computer Cartography*. Lund: Studentlitteratur, 1972.

RHIND, D. W., I. S. EVANS, and M. VISVALINGAM, "Making a National Atlas of Population by Computer," *Cartographic Journal*, 17, no. 1 (June 1980), 3–11.

ROSENFELD, AZRIEL, and AVINASH C. KAK, *Digital Picture Processing*. New York: Academic Press, Inc., 1978.

UNDERWOOD, S. H., and J. K. AGGARWAL, "Interactive Computer Analysis of Aerial Color Infrared Photographs," *Computer Graphics and Image Processing*, 6, no. 1 (February 1977), 1–24.

5 Vector Symbols

BAXTER, R. S., "Some Methodological Issues in Computer Drawn Maps," *Cartographic Journal*, 13, no. 2 (December 1976), 145–55.

BRASSEL, KURT E., and JACK J. UTANO, "Design Strategies for Continuous-tone Area Mapping," *American Cartographer*, 6, no. 1 (April 1979), 39–50.

CARLBOM, INGRID, and JOSEPH PACIOVEK, "Planar Geometric Projections and Viewing Transformations," *Computing Surveys*, 10, no. 4 (December 1978), 465–502.

CHASEN, SYLVAN H., *Geometric Principles and Procedures for Computer Graphic Applications*. Englewood Cliffs, N.J.: Prentice-Hall, Inc., 1978.

DAVIS, MICHAEL W. D., and MICHEL DAVID, "Generating Bicubic Spline Coefficients on a Large Regular Grid," *Computers and Geosciences*, 6, no. 1 (1980), 1–6.

FREEMAN, H., and J. A. SAGHRI, "Comparative Analysis of Line-Drawing Modeling Schemes," *Computer Graphics and Image Processing*, 12, no. 3 (March 1980), 203–23.

JEFFERY, MARGARET, HAZEL O'HARE, and CHRISTOPHER BOARD, "Choropleth Mapping on the Microfilm Plotter: An Attempt to Improve the Graphic Quality of Automated Maps," *International Yearbook of Cartography*, 15 (1975), 39–46.

KADMON, NAFTALI, "KOMPLOT 'Do-it-yourself' Computer Cartography," *Cartographic Journal*, 8, no. 2 (December 1971), 139–44.

MONMONIER, MARK S., "The Significance and Symbolization of Trend Direction," *Canadian Cartographer*, 15, no. 1 (June 1978), 35–49.

WAUGH, T. C., and D. R. F. TAYLOR, "GIMMS: An Example of an Operational System for Computer Cartography," *Canadian Cartographer*, 13, 2 (December 1976), 158–66.

WITTICK, ROBERT I., "A Computer System for Mapping and Analyzing Transportation Networks," *Southeastern Geographer*, 16, no. 1 (May 1976), 74–81.

YOELI, PINHAS, "Computer Executed Interpolation of Contours into Arrays of Randomly Distributed Height-points," *Cartographic Journal*, 14, no. 2 (December 1977), 103–8.

6 Cartometry and Map Projections

BARTON, B., "Note on the Transformation of Spherical Coordinates," *American Cartographer*, 3, no. 2 (October 1976), 161–68.

BOULTON, M. J. P., and C. C. TAM, "Computer Method for Radar Site Selection," *Cartographic Journal*, 11, no. 2 (December 1974), 117–20.

HILLIARD, JAMES A., UMIT BASOGLU, and PHILLIP C. MUEHRCKE, *A Projection Handbook*. Madison: Cartographic Laboratory, University of Wisconsin, 1978.

JENSEN, JOHN R., "Three-dimensional Choropleth Maps: Development and Aspects of Cartographic Communication," *Canadian Cartographer*, 15, no. 2 (December 1978), 123–41.

MOELLERING, HAROLD, "The Real-time Animation of Three-dimensional Maps," *American Cartographer*, 7, no. 1 (April 1980), 67–75.

MONMONIER, MARK S., "Nonlinear Reprojection to Reduce the Congestion of Symbols on Thematic Maps," *Canadian Cartographer*, 14, no. 1 (June 1977), 35–47.

OLSON, JUDY M., "Noncontiguous Area Cartograms," *Professional Geographer*, 28, no. 4 (November 1976), 371–80.

SUTHERLAND, IVAN E., ROBERT F. SPROULL, and ROBERT A. SCHUMACKER, "A Characterization of Ten Hidden-Surface Algorithms," *Computing Surveys*, 6, no. 1 (March 1974), 1–55.

TOBLER, W. R., "A Transformational View of Cartography," *American Cartographer*, 6, no. 2 (October 1979), 101–6.

———, "Analytical Cartography," *American Cartographer*, 3, no. 1 (April 1976), 21–31.

———, "Local Map Projections," *American Cartographer*, 1, no. 1 (April 1974), 51–62.

7 Cartographic Data Structures

AALDERS, H. J. G. L., "Data Base Elements for Geographic Information Systems," *ITC Journal*, no. 1980-1 (1980), pp. 76–85.

BABCOCK, HOMER C., "Automated Cartography Data Formats and Graphics: The ETL Experience," *American Cartographer*, 5, no. 1 (April 1978), 21–29.

BURTON, WARREN, "Representation of Many-Sided Polygons and Polygonal Lines for Rapid Processing," *Communications of the Association for Computing Machinery*, 20, no. 3 (March 1977), 166–71.

DUEKER, KENNETH J., "Land Resource Information Systems: A Review of Fifteen Years Experience," *Geo-Processing*, 1, no. 2 (December 1979), 105–28.

———, "Urban Geocoding," *Annals of the Association of American Geographers*, 64, no. 2 (June 1974), 318–25.

DYER, CHARLES R., AZRIEL ROSENFELD, and HANAN SAMET, "Region Representation: Boundary Codes from Quadtrees," *Communications of the Association for Computing Machinery*, 23, no. 3 (March 1980), 171–79.

FULTON, PATRICIA, and WILLARD L. McINTOSH, "Computerized Data Base for the Geomap Index," *American Cartographer*, 4, no. 1 (April 1977), 29–37.

MARTIN, JAMES, *Principles of Data-Base Management*, Englewood Cliffs, N.J.: Prentice-Hall, Inc., 1976.

MAURER, HERMANN A., *Data Structures and Programming Techniques*. Englewood Cliffs, N.J.: Prentice-Hall, Inc., 1977.

MONMONIER, MARK S., "Measures of Pattern Complexity for Choropleth Maps," *American Cartographer*, 1, no. 2 (October 1974), 159–69.

NAGY, GEORGE, and SHARAD WAGLE, "Computational Geometry and Geography," *Professional Geographer*, 32, no. 3 (August 1980), 343–54.

PAGE, E. S., and L. B. WILSON, *Information Representation and Manipulation in a Computer*, 2nd ed. London: Cambridge University Press, 1978.

PALMER, J. A. B., "Computer Science Aspects of the Mapping Problem," in John C. Davis and Michael J. McCullagh, eds., *Display and Analysis of Spatial Data*. London: John Wiley & Sons, Inc., 1975.

PEUCKER, THOMAS K., "Computer Cartography and the Structure of Its Algorithms," *World Cartography*, 15 (1979), 71–76.

———, "The Use of Computer Graphics for Displaying Data in Three Dimensions," *Cartographica*, 17, no. 2 (Summer 1980), 73–95.

———, and NICHOLAS CHRISMAN, "Cartographic Data Structures," *American Cartographer*, 2, no. 1 (April 1975), 55–69.

SAMET, HANAN, "Region Representation: Quadtrees from Boundary Codes," *Communications of the Association for Computing Machinery*, 23, no. 3 (March 1980), 163–70.

SANDBERG, G., "A Primer on Relational Data Base Concepts," *IBM Systems Journal*, 20, no. 1 (1981), 23–40.

SHAPIRO, LINDA G., "Data Structures for Picture Processing: A Survey," *Computer Graphics and Image Processing*, 11, no. 2 (October 1979), 162–84.

TOMLINSON, R. F., H. W. CALKINS, and D. F. MARBLE, *Computer Handling of Geographical Data: An Examination of Selected Information Systems*. Paris: UNESCO Press, 1976.

WERNER, P. A., "National Geocoding," *Annals of the Association of American Geographers*, 64, no. 2 (June 1974), 310–17.

WILLIAMS, ROBIN, "A Survey of Data Structures for Computer Graphics Systems," *Computing Surveys*, 3, no. 1 (March 1971), 1–21.

8 Computer-Assisted Map Design

BOYLE, A. R., "The Quantized Line," *Cartographic Journal*, 7, no. 2 (December 1970), 91–94.

BRYCH, DAVID J., "Queuing Theory and CPM/PERT Applications for Effective Cartographic Production Management," *Proceedings*, ACSM Fall Technical Meeting, 1978, 39–52.

DOUGLAS, DAVID H., and THOMAS K. PEUCKER, "Algorithms for the Reduction of the Number of Points Required to Represent a Digitized Line or Its Caricature," *Canadian Cartographer*, 10, no. 3 (December 1973), 112–22.

DUTTON, GEOFFERY H., "Fractal Enhancement of Cartographic Line Detail," *American Cartographer*, 8, no. 1 (April 1981), 23–40.

EVANS, IAN S., "The Selection of Class Intervals," *Transactions of the Institute of British Geographers*, new series, 2, no. 1 (1977), 98–124.

GUPTILL, STEPHEN C., "An Evaluation Technique for Categorical Maps," *Geographical Analysis*, 10, no. 3 (July 1978), 248–61.

JENKS, GEORGE F., "Lines, Computers, and Human Frailties," *Annals of the Association of American Geographers*, 71, no. 1 (March 1981), 1–10.

——— and FRED C. CASPALL, "Error on Choropleth Maps: Definition, Measurement, Reduction," *Annals of the Association of American Geographers*, 61, no. 2 (June 1971), 217–44.

MARINO, JILL S., "Identification of Characteristic Points along Naturally Occurring Lines: an Empirical Study," *Canadian Cartographer*, 16, no. 1 (June 1979), 70–80.

MONMONIER, MARK S., "The Hopeless Pursuit of Purification in Cartographic Communication: A Comparison of Graphic-arts and Perceptual Distortions of Graytone Symbols," *Cartographica*, 17, no. 1 (June 1980), 24–39.

———, "Viewing Azimuth and Map Clarity," *Annals of the Association of American Geographers*, 68, no. 2 (June 1978), 180–95.

MORRISON, JOEL L., "A Theoretical Framework for Cartographic Generalization with Emphasis on the Process of Symbolization," *International Yearbook of Cartography*, 14 (1974), 115–27.

RAUHALA, URHO A., "Introduction to Array Algebra," *Photogrammetric Engineering and Remote Sensing*, 4, no. 2 (February 1980), 177–92.

RHIND, D. W., "Generalisation and Realism within Automated Cartographic Systems," *Canadian Cartographer*, 10, no. 1 (June 1973), 51–62.

VANICEK, P., and D. F. WOOLNOUGH, "Reduction of Linear Cartographic Data Based on Generation of Pseudo-Hyperbolae," *Cartographic Journal*, 12, no. 2 (December 1975), 112–19.

WILLIAMS, ANTHONY V., "Interactive Cartogram Production on a Microprocessor Graphics System," *Proceedings*, ACSM Fall Technical Meeting, 1978, 426–31.

YOELI, PINHAS, "The Logic of Automated Map Lettering," *Cartographic Journal*, 9, no. 2 (December 1972), 99–108.

Sources of Additional Information

A variety of government agencies, private institutions, and professional organizations promote the development of mapping software and the exchange of information about mapping and automated data processing. This appendix describes briefly some of the more prominent sources to which the reader may write for additional information about current developments in various aspects of computer-assisted cartography.

American Congress on Surveying and Mapping

Established in 1941, the American Congress on Surveying and Mapping (ACSM) is a professional organization serving cartographers, geodesists, and land surveyors. ACSM was reorganized in 1981, as a federation of member organizations, one of which is the American Cartographic Association (ACA). Cartographers belong to both ACSM and ACA, and members receive all ACSM publications including *The American Cartographer,* a semiannual journal with articles and reviews on traditional and computer-assisted cartography. Full-time students pay a substantially reduced membership fee. ACSM is a member of the International Cartographic Association (ICA), which publishes occasional monographs on computer-assisted cartography. Among ICA-sponsored publications available through ACSM are the *Glossary of Terms in Computer-Assisted Cartography* and the annual *International Yearbook of Cartography.* At the annual spring and fall technical meetings, held jointly with the American Society of Photogrammetry, technical paper sessions, workshops, and a large exhibit provide an opportunity to keep abreast of current developments in all phases of mapping and remote sensing.

For additional information, write to:

> American Congress on Surveying and Mapping
> 210 Little Falls Street
> Falls Church, Virginia 22046

Geo Abstracts

Geo Abstracts is a useful bibliographic aid published in seven parts, each with six issues a year. Computer-assisted cartography is covered in Part G, Remote Sensing, Photogrammetry, and Cartography. Approximately 2,500 articles, books, and technical reports are abstracted annually, with about 20 percent listed under the more directly relevant headings "Data processing," "Data acquisition—theory," and "Automated mapping." Although some of the one-paragraph summaries are taken directly from the abstracts accompanying journal articles or are otherwise provided by authors, many abstracts are written expressly for *Geo Abstracts* by subject-matter experts or are borrowed from other collections of abstracts. Annual author and regional indexes are included in the last issue of each volume, and an annual combined index for Parts A, B, E, and G (covering topics in physical geography, in contrast to human geography) includes a permuted index of key words in the titles.

For additional information, write to:

> Geo Abstracts Ltd.,
> University of East Anglia,
> Norwich, NR4 7TJ,
> England

Geography Program Exchange

The Geography Program Exchange (GPE) was established in 1971 to assist colleges, universities, and other nonprofit organizations with the exchange and acquisition of computer software. GPE maintains a central file of computer programs for mapping and geographic analysis. No attempt is made to duplicate standard programs for statistical analysis. GPE distributes programs, together with associated documentation and test data sets, at cost on a nonroyalty basis. A catalog is available, listing over 150 programs, approximately 25 percent of which are directly concerned with mapping.

For additional information, write to:

> Geography Program Exchange
> Department of Geography
> 315 Natural Science Building
> Michigan State University
> East Lansing, Michigan 48824

Harvard University Laboratory for Computer Graphics and Spatial Analysis

Established in 1965, the Harvard Laboratory is a principal center for the development of mapping software and the dissemination of information about computer graphics to business, government, and higher education. The Laboratory publishes the informative twice-monthly *Harvard Newsletter on Computer Graphics,* the growing collection of research and applications papers in the *Harvard Library of Computer Graphics Mapping Collection,* and the annual *Directory of Computer Graphics Suppliers.* Several widely used batch and interactive mapping programs, such as SYMAP, CALFORM, SYMVU, POLYVRT, ASPEX, and DOT.MAP, the sophisticated ODYSSEY data analysis and display system, and a number of coordinate data bases are described in the Laboratory's periodic catalog, *Lab-Log.* Considering the time required to write and test one's own software, Harvard's prices are nominal. The Laboratory also conducts workshops and seminars in various regional centers throughout the United States and in several foreign countries, and hosts the annual Harvard Computer Graphics Week in midsummer.

For additional information, write to:

> The Laboratory for Computer Graphics and Spatial Analysis
> Harvard University
> 520 Gund Hall
> 48 Quincy Street
> Cambridge, Massachusetts 02138

National Technical Information Service

The National Technical Information Service (NTIS) of the U.S. Department of Commerce is the central source for the public sale of government-sponsored research, development and engineering reports, and for federally generated machine-processable data files. NTIS can provide copies of mapping programs such as CAM (the Cartographic Automatic Mapping system developed by the Central Intelligence Agency) and geographic data bases such as World Data Banks I and II and a national county-boundary file. Every two weeks NTIS publishes *Government Reports: Announcements and Index,* with abstracts of federally supported research reports. A number of titles relevant to mapping and map analysis appear in each issue, in Group 8B—Cartography. These reports are sold as paper copy, microfiche, or microfilm, and computer software is distributed on magnetic tape. NTIS has more than 30 cooperating organizations around the world, which provide local access to its products and services.

For additional information, write to:

> Computer Products
> NTIS
> Springfield, Virginia 22161

SIGGRAPH/Association for Computing Machinery

SIGGRAPH is the Special Interest Group on Graphics of the Association for Computing Machinery (ACM). SIGGRAPH holds an annual technical conference and publishes a quarterly report, *Computer Graphics,* with articles on all aspects of graphical person-machine communication, including hardware, language and data structures, methodology, and applications. Persons whose major professional allegiance is in a field other than information processing may join SIGGRAPH without joining ACM.

In addition to over 30 Special Interest Groups (SIGs), ACM also sponsors numerous local chapters. ACM holds national and regional meetings for the presentation and discussion of papers and for demonstrations by hardware and software exhibitors, and has an extensive publications program. Among the purposes of ACM are the advancement of the science and art of information processing and the free interchange of information among both specialists and the public.

For additional information, write to:

Association for Computing Machinery
1133 Avenue of the Americas
New York, New York 10036

U.S. Bureau of the Census

The Bureau of the Census is the principal agency in the United States for the collection and distribution of social and economic data. Digital data files available from the Bureau include DIME files describing the street networks and enumeration units of major urban areas and summary tapes for the Census of Population and Housing, the Census of Agriculture, and the other economic censuses. Particularly useful for business analysts, social scientists, and planners is the Historical State, County, and City Data File, with a variety of data recorded for these areal units at intervals of approximately 5 years beginning in the mid-1940s. This extensively documented file is available on two computer tapes. The Bureau also provides individual tape files for the *County and City Data Book,* a standard reference work published every 5 years.

For additional information, write to:

Customer Services Branch
Data User Services Division
Bureau of the Census
Washington, D.C. 20233

U.S. Geological Survey

The U.S. Geological Survey is responsible for the National Mapping Program and is the principal agency for large-scale base mapping. A variety of digital data bases are available, including digital elevation models (DEMs) and digital line graphs (DLGs) corresponding to the Survey's large-scale, 7.5-minute quadrangle maps; land-use and land-cover data for analysis and plotting at 1:250,000; the 1:2,000,000 National Atlas sectional map files; and a small-scale, special-subject thematic data base for physical features and environmental phenomena, suitable for analysis and plotting at 1:7,500,000. Information on current digital products and projects, as well as information about satellite imagery, can be obtained from the Survey's National Cartographic and Geographic Information Service. Geological Survey employees participate actively in the technical program of the American Congress on Surveying and Mapping, and USGS has collaborated with the Bureau of the Census and ACSM in sponsoring several International Symposia on Computer-Assisted Cartography, a series of widely attended conferences better known by the acronym AUTO-CARTO.

For additional information, write to:

National Cartographic and Geographic Information Service
U.S. Geological Survey
507 National Center
Reston, Virginia 22092

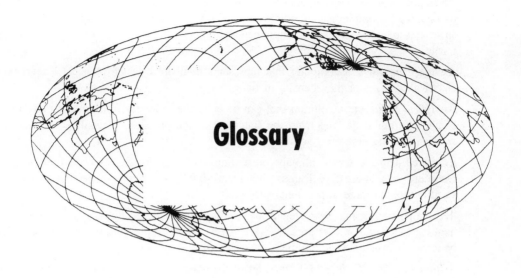

Glossary

Terms are defined here only as used in this book. Rapid advances in computer technology have strained the ability of language to provide concise, widely recognized descriptors of new machines and concepts. The development of jargon is inevitable in these circumstances, as is the confusion produced by a fluid vocabulary of ephemeral words used differently by various authors and speakers. This glossary is intended principally to serve the reader of this book and thus includes possibly unfamiliar terms in cartography and mathematics, as well as words and phrases related to computers and data structures.

Acoustic Coupler: an audio-digital translating device for converting a terminal's digital signals to chirps sent over a telephone line, and for converting the chirps received to digital signals intelligible to the computer.

Acronym: a word derived from the first letter or first few letters of a series of words, for example, CRT for cathode-ray tube.

Address: a number referring to a location in computer memory.

Address Coding Guide (ACG): a geographic base file that relates street addresses to census blocks, tracts, or other administrative or enumeration units.

Algorithm: a series of specific steps for solving a problem.

Alphanumeric: consisting of both letters and numbers, and possibly including other symbols such as punctuation marks.

Anaglyph: a stereoscopic diagram in which the two views are printed or projected superimposed in complementary colors, red and blue or red and green. By viewing through spectacles of corresponding complementary colors, the user can form a stereoscopic image.

Analog Computer: a computer that represents numbers by such physical quantities as electrical voltages or the intensity of light.

Analytical Solution: a solution that can be calculated directly from one or more mathematical formulas derived from the statement of the problem. In contrast, an *iterative solution* is a trial-and-error solution.

Animated Map: a series of maps presenting a moving image, as on a film or video recording with successive frames shot from slightly different perspective views of a steadily rotating three-dimensional surface.

Arcs: a portion of the perimeter of a two-dimensional closed figure lying between two nodes at which two or more arcs intersect. An arc usually represents a continuous common boundary between two adjoining mapping units.

Arithmetic Logic Unit (ALU): the part of the computer that performs arithmetic calculations and such logical tests as "greater than" and "equal to."

Array Processor: a computer designed to operate simultaneously, or "in parallel," on many elements of an array.

Aspect: the horizontal direction in which a slope faces, commonly expressed in degrees clockwise from north. A west-facing slope, for example, would have an aspect of 270 degrees.

Assembler: a program for translating instructions written in a lower-level language into machine instructions.

Associative Storage: the representation of data items in memory with codes or *keys* so that items specified by a key can be retrieved without specification of their memory locations; also called *content-addressable memory*.

Azimuth: the horizontal direction measured clockwise from north. "Due east," for example, has the azimuth 90 degrees.

Band: a range of wavelengths.

Batch Processing: the collection of programs in a queue for later processing, perhaps with priorities assigned according to estimates of the processing time required.

Bearing: the horizontal angle of a line of direction, measured in the quadrant of the line as degrees east or west of the meridian of reference. Northeast, for example, has a bearing of 45 degrees east of north, whereas southwest has a bearing 45 degrees west of south.

Bit: abbreviation for *bi*nary digi*t*; a number that can take only the values 0 or 1.

Bit Plane: a part of a data structure of superimposed grids of cells having the values 0 or 1.

Block: data records stored near each other for reference and processing as a group.

Boundary Segments: shared continuous boundaries between adjoining areal units.

Bug: an error in a program or computer system.

Bulk Storage Unit: a supplementary memory unit with a somewhat slower access time than the main memory of a computer; also called a *backing store*.

Byte: a group of bits that can be stored and retrieved as a unit.

Cadastre: a survey that creates, defines, retraces, or reestablishes the boundaries and subdivisions of public lands and private estates. The ownership and value of private lands are recorded for taxation.

Canned Program: a program prepared for widespread distribution, usually with user-oriented documentation and sufficient flexibility in data entry so that individual users do not need to alter the program.

Cartogram: a map projection with areas or distances distorted, according to a *transforming variable,* to communicate by relative distance or size such concepts as travel time or population size.

Ceiling Function: a mathematical function that rounds a decimal number upward to the next highest integer.

Chain: an ordered arrangement of items, as in a list of data items with each record linked to both the previous and following records in the case of a *two-way* chain and only to the following record in the case of a *one-way* chain.

Chip: a small piece of silicon containing the reproduced image of an *integrated circuit.*

Chord: a straight line joining any two points on an arc, curve, or circular circumference.

Choropleth Map: a map with areas colored or shaded so that the darkness or lightness of an area symbol is proportional to the density of the mapped phenomenon.

Chroma: degree of saturation or purity of a color.

Clipping: the severing at the frame of a line segment lying partly outside a *window* to be displayed on a CRT unit or plotter.

Color Composite: in satellite imagery, a color negative, transparency, or print produced by allowing the reflectances recorded for each band of a multispectral image to be represented by a proportionate intensity of one of the primary colors.

Color Separations: separate drawings prepared for each type of map data to be reproduced in a separate color.

COM Unit: an acronym for computer output on microfilm; a device that plots graphic images on film with a narrow beam of radiant energy. COM images are usually enlarged before printing.

Compiler: a program for translating instructions written in a high-level computer language into machine instructions.

Computed Address: a numeric representation of a memory location computed from several variables such as the starting address and dimensions of a two-dimensional array and the subscripts for a specific element in the array.

Computer Architecture: the design of a computer system's memory, peripherals, central processor, and especially the circuitry for control, logic, storage, and retrieval.

Constant: a word in memory identified by a label in the program and, in contrast to a *variable,* assigned a value that generally does not change during the execution of the program.

Continuous Mode: the digitizing of a series of points along a linear feature tracked by the operator with point locations recorded automatically at a constant time interval.

Contour: an imaginary line all points of which are at the same elevation.

Contour Threading: symbolization of a three-dimensional surface by the serial estimation of points at which a contour passes between successive sets of control points. A contour will pass between two control points if one has a value above and the other has a value below the value of the contour.

Control Point: a point with a given horizontal position and a known surface elevation to be used in estimating unknown elevations elsewhere in the area to be mapped.

Control Unit: the part of the computer that directs the sequence of operations, interprets the digitally coded machine instructions, and initiates the proper signals to the computer circuits to execute the instructions; also called a *monitor.*

Core: (1) a tiny magnetic ring, the polarity of which can be varied to store a bit as either 0 or 1; (2) the main memory of a computer.

Cosine: the ratio between the side adjacent to a specified acute angle in a right triangle and the hypotenuse.

CPU: an acronym for central processing unit, the part of the computer that controls the flow of data and performs the computations.

CRT: a cathode-ray tube, similar to a television picture tube, on which an image is displayed by a pattern of glowing spots produced by directing a beam of electrons at a phosphorescent screen.

Cursor: the hand-held, movable part of a digitizer with cross hairs for the accurate designation of points on an image.

Data Structure: the organization in memory of data, and, in particular, the reference linkages among data elements.

Debug: to detect and correct an error in a program or computer system.

Declaration: a program statement indicating the dimensions of an array, whether a variable is to contain integer, floating point, or character data, or some other specification related to the assignment of memory to the program.

Dedicated System: a computer system allocated to a single specified use.

Density Slicing: the classification of digital multispectral remotely sensed images by assigning to different categories pixels for which the recorded reflectances in a band lie on different sides of a "slice level."

Deque: a linear list data structure with two ends at which data items can be either stored or retrieved.

Determinant: a single value computed from a two-dimensional array of numbers having an equal number of rows and columns. The determinant of a 2 by 2 array is computed by subtracting the product of the off-diagonal elements from the product of the diagonal elements.

Digital Computer: a computer that represents numbers by counters, usually in binary digits.

Digital Elevation Model (DEM): a file with terrain elevations recorded for the intersections of a fine-grained grid, commonly obtained with an orthophotoscope and organized by quadrangle as the digital equivalent of the elevation data on a topographic base map.

Digital Line Graph (DLG): a file containing lists of point coordinates describing boundaries, drainage lines, transportation routes, and other linear features, commonly digitized manually or scanned and organized by quadrangle as the digital equivalent of the linear hydrographic and cultural data on a topographic base map.

Digital Tablet: a device used to determine and communicate to a computer the coordinates of points designated with a *stylus*. Locations are sensed electronically by the stylus and a tablet bed on which a graphic image or *instruction menu* is placed.

Digital Terrain Model (DTM): a land surface represented in digital form by an elevation grid or lists of three-dimensional coordinates.

Digitizer: a device for converting point locations on a graphic image to plane (x, y) coordinates for digital processing.

DIME File: a geographic base file with *D*ual *I*ndependent *M*ap *E*ncoding, with the linear elements in a network coded to represent both their *bounding* by areal units and their *cobounding* by the points at which the linear links meet.

Direct Access: referring to a peripheral memory unit or retrieval process in which the time required to retrieve an item of data is independent of the location of the item.

Directed Graph: a graphic representation of a data structure in which lines represent only one-way linkages between data items.

Directional Isolation: the character of a data point located away from other data points on the opposite side of an interpolated point.

Directory: a look-up table indicating the storage locations in a file of various data records and used for gaining access to these records.

Distributed Data Base: a data base with unique components in geographically dispersed locations linked through a telecommunications network.

Documentation: the written specifications of a program indicating the program's goals, memory requirements, data structures, and algorithms; the description and format for the data to be entered; and the description and format of the results.

Dot-distribution Map: a map showing geographic variations in density with dot symbols, each of which usually represents a constant, specified amount of the mapped phenomenon.

Double Precision: the use of two computer words to represent a quantity with more significant digits than in a single-precision word.

Drum Plotter: a device with a rotating cylindrical drawing surface and paper reels for plotting graphic images on a long roll of paper.

Drum Scanner: a device for measuring and recording reflectances from an image mounted on a cylinder rotating beneath a scan head, which migrates slowly along the length of the cylinder in order to cover the entire image.

Edge: a line linking two nodes in a graphic representation of a data structure; also called a *side* or an *arc* in graph theory.

Edge Growth: the enlargement during photography, plate etching, or printing of the elements of an image, especially the dots in a tint screen or the fine lines of a line screen. Edge growth produces darker area symbols than intended.

Edge Matching: the comparison and graphic adjustment of features to obtain agreement along the edges of adjoining mapped regions.

Electromagnetic Spectrum: the ordered array of known electromagnetic radiations, extending from the shortest cosmic rays, gamma rays, and x-rays, through ultraviolet radiation, visible radiation, and infrared radiation, and including microwave and radio wavelengths.

Electromechanical: a mechanical device, such as certain types of digitizer, the operation of which depends upon electric circuitry, sensors, or motors.

Electrostatic Printer: a device for printing graphic images by placing small electrical charges on the paper so that a dark or colored powder, or *toner,* will adhere in these spots.

Error Message: a code or diagnostic message displayed when the computer detects a mistake or inconsistency while a program is being compiled, interpreted, or executed.

External Memory: memory separate from the main memory of a computer but holding information to be read or saved by the computer.

Extrapolation: the estimation of surface elevations in areas beyond those with data.

Eye Base: the distance between the centers of the pupils.

Factor Analysis: a multivariate technique for reducing a set of somewhat redundant measurements to a smaller number of significant, nonredundant measurements.

Feature Code: a number representing a category of feature, for example, 280.7020 to indicate a pipeline.

Fetch Function: a function that retrieves data from storage for use by the central processor.

File: a collection of records, each of which can be referenced according to its position in the file.

Finite-state Machine: the concept in the theory of automatic machines of a machine that can have only a finite number of states. This concept is useful in developing theorems to prove whether a task can be carried out by a digital computer.

Firmware: software available for distribution as hardware, for example, a program or data file stored permanently on a silicon chip or plug-in ROM.

Flatbed Plotter: a device with a flat drawing surface for plotting a graphic image from a list of point coordinates and pen codes.

Floating Point: a notation in which a number x is represented by a *mantissa y* and an integer exponent z so that $x = y \cdot n^z$, where n is usually the integer 10.

Floor Function: a mathematical function that rounds a decimal number downward to the next lowest integer.

Floppy Disk: a circular, flexible, relatively inexpensive piece of magnetic material for the storage of digital data.

Flow Chart: a graphic representation of an algorithm, with well-defined steps linked by flow lines to indicate the sequence of operations.

Font: an assortment of letters, numbers, and special characters of the same typeface.

Foreshortening: the representation of some lines relatively shorter than other lines in order to achieve the illusion of a three-dimensional object in a perspective drawing.

Fragmentation Index: a numerical measure of the pattern complexity of a choropleth map, as indicated by the relative frequency of spatially separate groups of areal units represented by the same symbol.

Generable Behavior: a pattern of responses normally ascribed to human beings but which can be simulated by a computer.

Geographic Direction: direction measured relative to a north meridian, in contrast to direction based upon magnetic north or grid north.

Glitch: a relatively minor but possibly annoying mistake, for example, an unexplained erratic jog in a digitized line.

Global Optimum: a solution to an optimization problem yielding a value of the objective function higher or lower than that provided by any other possible solution.

Gnomonic Projection: a projection of the sphere onto a plane, with every great circle represented by a straight line.

Graduated Circle: a circular symbol with its area proportional to the magnitude represented.

Graticule: the network of parallels of latitude and meridians of longitude as plotted on a map projection.

Gray Scale: an ordering of shades of gray from white to black.

Great Circle: the line of intersection of the surface of a sphere with a plane passing through the center of the sphere. A great circle is the largest circle that can be drawn on a sphere.

Hachures: a series of lines representing the general direction and steepness of slope. Gentle slopes are portrayed by longer, lighter, or more widely spaced lines, whereas steep slopes are symbolized by shorter, heavier, or more closely spaced lines.

Halftone Dots: the small, uniform, evenly spaced circular or squarish elements of a contact screen used in graphic arts for converting a continuous-tone photograph into a finely patterned, printable image with tonal contrasts.

Hard Copy: a permanent image of a map or diagram, for example, a paper map produced on a line printer or pen plotter.

Hardware: electronic and mechanical equipment for computing, graphic display, data storage, and data transmission.

Hash Coding: a technique used to conserve memory when a file otherwise would have a large number of unused records. *Hashing functions* are used to convert hypothetical positions in a potentially wasteful file into real positions in a more compact file in which the data items are actually stored.

Hidden-line Algorithm: a programmable procedure for removing hidden portions of the projected profiles representing a three-dimensional surface.

Host System: the computer system supporting a terminal, usually in time share through a telecommunications network.

Hue: that dimension of color related to wavelength and the classification of the color as, say, a blue, a green, or a red.

Infrared: the portion of the electromagnetic spectrum bounded by visible light and by microwave radiation. Infrared wavelengths just longer than those of visible light are called *near infrared,* whereas the longer infrared wavelengths, sensible as heat, are called *thermal infrared.*

Initialize: to set program variables to their starting values, commonly zero, at the beginning of a program.

Ink-jet Printer: a display device that prints characters and graytones as patterns of small dots formed by tiny drops of ink sprayed onto the plotting medium.

Integrated Circuit: a small piece of silicon the surface of which contains a microscopic reproduced image that is the electronic equivalent of thousands of transistors.

Interactive Graphics System: a computer system with a relatively rapid response to the instructions of a user entering, editing, manipulating, or displaying graphic data.

Interface: an electronic translator of the signals of two devices, such as a computer and a plotter, so that otherwise incompatible information can be transferred between them.

Internal Memory: that integral part of a computer immediately accessible by the CPU for the storage and retrieval of data, also called *main memory* or *core storage*.

Interpolation: estimation of surface elevations in areas with data values nearby, on more than one side.

Interpreter: a program that translates a source program written in a high-level programming language to machine code, instruction by instruction, as the source program is being processed or as the operator of an interactive system enters instructions.

Intersection: the set containing all objects common to two intersecting sets.

Island: a closed two-dimensional figure. In GIRAS, the Geographic Information Retrieval and Analysis System developed at the U.S. Geological Survey, an island is a unit of land cover lying completely within another land-cover unit.

Isoline Map: a map with the form of a surface shown by lines connecting points of equal value. A contour map is also an isoline map.

Isopleth Map: a map with lines connecting places estimated to have equal values for the mapped distribution.

Iteration: the single execution of a set of instructions programmed for repetition in a *loop*.

Joystick: a lever with two or more degrees of freedom that is used to indicate point coordinates with a target image positioned on a CRT screen by the operator.

Key: a digital code used to identify an item of information, and which is used in a retrieval to accept or reject items for further processing.

Kilobyte: a unit of memory representing 1,024 bytes and often designated with the symbol K, as in 4K for 4 kilobytes. The symbol K is also used to refer to 1,024 words of any specified size.

Kriging: an interpolation procedure for obtaining statistically unbiased estimates of surface elevations from a set of control points.

Laser-beam Plotter: a device that plots graphic images on photosensitive film with a thin beam of coherent light, which can be focused accurately to produce a sharp image.

Least-squares: a method of estimation for which the best result is that yielding the minimum value for the sum of the squared deviations between, in the case of multiple regression for a trend surface, the actual and trend values.

Light Pen: a pointing device used to indicate positions on the screen of a CRT unit so that images and parts of images can be drawn, deleted, or moved.

Line Printer: a display device for printing alphabetic, numeric, and other symbols line by line on sheets of paper, usually joined along a perforated edge.

Linear Discriminant Function: a linear algebraic expression for calculating values useful in the multivariate technique *discriminant analysis* for classifying places and other types of data unit.

Linear Interpolation: the estimation of a surface elevation with the value at each data point weighted according to the inverse of its distance from the interpolated point.

Linear List: the ordering of data items in a sequential structure with each item linked only to the items immediately preceding and following it.

List Structure: the ordering of data items with each item linked to one or more items in the list.

Lithography: a printing process in which the image to be printed is ink receptive and all other areas are ink repellent. In *offset lithography* the inked image is first transferred from a flat or cylindrical press plate to a rubber-surfaced cylinder and then printed, or offset, onto a sheet or a continuous web of paper.

Local Operator: a computational and sampling process that employs only nearby data values in determining such spatial properties as slope and the elevation at a point of a smoothed surface.

Local Optimum: a solution to an optimization problem yielding a value of the objective function better than that for any slightly different solution but not better than at least one other, more distinctly different solution.

Logical Operation: an operation involving program variables that is either true or false. Three examples of logical operations are "*A* equal to *B*," "*C* greater than *D*," and "*A* equal to *B* AND *C* greater than *D*."

Loop: repetition of a group of instructions in a program.

LSI: large-scale integration, whereby such complex circuits as a CPU or large amounts of memory can be contained in one integrated circuit. An even more advanced technology is referred to as VLSI, for very large scale integration.

Machine Language: instructions and declarations in a digital code that can be acted upon by the computer directly, without further translation.

Magnetic Disk Unit: a peripheral device with rotating circular magnetic plates for storing data.

Magnetic Drum Unit: a peripheral device with a rotating magnetically coated cylinder for storing data.

Magnetic Tape: a long, thin, magnetically coated storage medium, usually wound on a reel.

Mantissa: the decimal part of a floating point number, as distinguished from the *exponent*.

Mask: a sheet of film or other material placed between a film image and a photosensitive sheet of film in order to block the transfer of parts of the film image.

Matrix Printer: a display device that forms characters and prints graytones with a pattern of dots from, say, a 10 by 10 grid of available dots for each character position.

Memory: an organized set of locations in which a computer can store and find data and instructions.

Menu: a list of options on a display allowing an operator to select the next operations by indicating one or more choices with a pointing device, such as the stylus of a digital tablet.

Mercator Projection: a map projection centered along the Equator with evenly spaced meridians perpendicular to parallels spaced progressively farther apart poleward so that compass bearings are not distorted. A *transverse Mercator projection* uses the same system of projection, but with the projection centered along a meridian to provide low distortion with a zone around the *central meridian*.

Microfilm Recorder: a device that plots graphic images on film with a narrow beam of radiant energy.

Microprocessor: a computer occupying comparatively little space through the use of integrated circuitry.

Minicomputer: a comparatively inexpensive computer limited in internal memory and often in word size.

Mnemonic: a code intended to promote recall from the human memory.

Modem: a translating device that links a terminal to a telecommunications network. An *acoustic coupler* is a modem that permits a terminal to communicate through the handset of a standard telephone instrument.

Moebius Strip: a surface with only one side, as might be formed by joining, after a half-twist, the opposite ends of a long strip of paper.

Moiré Effect: a noticeable pattern of wavy, regularly spaced spots resulting from the nearly parallel alignment of the axes of two superimposed dot screens.

Mosaic: an assembly of aerial photographs or maps matched along the edges to provide a continuous representation of part of the earth's surface.

Multispectral Scanner (MSS): a device that oscillates a flat mirror between the field of view and a set of optical sensors in order to gather data on several bands simultaneously.

Node: a point at which two or more lines meet; called an *edge* or a *vertex* in graph theory.

Objective Function: the mathematical expression that is to attain a maximum or minimum value in an optimization process. The range of *feasible solutions* is usually limited by a set of *constraints*.

Off-line: the transmission of information between a computer and a peripheral unit before or after, but not during, processing, in contrast to *on-line* processing.

On-line: the transmission of information between a computer and a terminal or display device while processing is occurring, in contrast to *off-line* processing.

Open Window Negative: a film negative with open areas, used as a mask where screens are to be printed in the open areas.

Optical-fiber Transmission: the rapid and relatively distortion-free transmission of information through fine optical fibers, with digital data transmitted as a series of on-off pulses of light representing bits of information.

Orthophotoscope: an instrument for scanning a projected stereomodel and recording a new photographic image with accurate horizontal distances.

Overflow: a condition in which a value is too large to be represented in a word in memory, for example, when a program attempts to store the computed result 0.1×10^{156} in a word not able to accept a positive exponent larger than 127.

Page: a segment of a program or its associated data stored in either internal or external memory, as needed, on a computer having *virtual memory*.

Page Thrashing: the inefficient operation of a computer having *virtual memory* when an incompatibility between data structure and algorithm results in the computer devoting an inordinately large amount of time to swapping *pages* between internal and external memory.

Pallet: the selection of standard colors available to the user of an interactive color graphics system.

Parallax: the apparent displacement of the position of an object with respect to a reference point caused by a shift in the point of observation. The *parallax difference,* the difference in displacement of two points as portrayed on two overlapping aerial photographs, can be used to compute the difference in elevation.

Parameter: a constant in a mathematical expression that varies to indicate a particular state or circumstance. For example, the radius r of a circle expressed as $Y = (r^2 - x^2)^{1/2}$.

Parcel Block Inventory: a geographic base file for land parcels containing such information as a description of boundaries, the name of the owner, and the assessed value.

Parity: a number assigned to a series of grid cells to indicate membership in the same linear feature or land-cover type.

Parse: to scan a list or sequential file of uncertain length to determine its contents.

Path: a line or course of direction through a tree data structure from the *root* to a lower-level data element.

Peelcoat: a transparent plastic material with an opaque coating that can be etched or cut and then peeled away to prepare open windows for printing masks.

Pen Code: a code specifying whether a plotter is to lift or lower the pen in moving to a new location on the drawing.

Penalty Function: a mathematical expression incorporated in the objective function of an optimization process to avoid, but not rule out, solutions with unwanted characteristics.

Peripheral: a device that may be added to a computer to provide additional data storage or to receive or display data.

Perspective Projection: the projection of points by straight lines through a single, given point to their intersections with the plane of projection.

Photogrammetry: the science or art of obtaining reliable measurements by means of photographs.

Photohead: a drawing unit that uses a light beam to plot a graphic image on photosensitive film; sometimes called a *lighthead*. A photohead is usually designed for use with a flatbed plotter, which shifts it from point to point on the drawing.

Photomultiplier Tube: an electron-emitting, light-sensitive vacuum tube in which the intensity of the light energy received controls the flow of electrons in order to yield an amplified signal, as in a sensor of visible light.

Pixel: an element of a digital map or picture representing either (1) a small square or nearly square portion of the Earth's surface sensed from a satellite or aircraft scanner or (2) a square portion of a graphic image sensed by an optical scanner; also called a *pel*, for *pi*cture *el*ement.

Plane Coordinates: coordinates specifying the locations of points in a plane. In cartography the plane usually is a projection of the Earth's surface onto a flattenable cone or cylinder, and the X and Y values scaled along the rectangular axes are called *eastings* and *northings*, respectively.

Planimetric Map: a map that represents only the horizontal positions of features. A *planimetrically accurate* map shows accurate horizontal distances between features.

Platen: the flat surface of a digitizer, digital tablet or flatbed plotter, or the curved surface of a drum plotter or drum scanner.

Plex: a complex data structure with items linked as a network rather than as a simpler linear list or tree.

Point Mode: the digitizing only of points specifically indicated by the operator.

Pointer: a numerical value that designates directly or after some calculations the next data element in a list or a cross-reference to some related data element.

Polygon: a two-dimensional figure with three or more straight-line sides intersecting at a like number of points. *Area polygon* may refer to an enumeration district or irregularly shaped land parcel represented by a list of point coordinates.

Polynomial: a linear combination of the products of integer powers of a set of variables, for example, $X^2 + 5X + 4XY$.

Press Plate: a thin metal plate containing a printable image from a color-separation drawing.

Primary Colors: basic colors from which other hues can be produced. The *additive* primaries (red, green, and blue) are used with light, whereas the *subtractive* primaries (yellow, magenta, and cyan) are used with printing inks.

Print Train: a closed loop of type slugs used on a high-speed printer. With some line printers the print train, or *print chain,* may be exchanged to provide other sets of characters.

Program: a set of declarations and logically organized instructions coded in a computer language in order to direct the operation of the computer.

Programming Language: a formal set of verbal or symbolic instructions and declarations that can be used to code an algorithm for later translation into machine instructions.

Proximal Map: a surface map with area symbols assigned so that the symbol anywhere on the map represents the data value at the closest data point.

Queue: a linear list data structure with one "joining" end at which data items are stored and one "leaving" end at which data items are retrieved.

Random-access File: a file, such as a disk or drum file, for which the time required for access is independent of the location of the information most recently stored or retrieved.

Raster: the pattern of horizontal, parallel scan lines comprising the image on a CRT screen, on which each scan line consists of segments varying in intensity. *Raster data* thus refers to data in the form of parallel scan-line segments or grid cells.

Record: a group of items in a file treated as a unit. For example, all data items for a census tract can be grouped as a record and assigned to a single segment of a magnetic tape file for convenient storage and retrieval.

Rectification: the process of removing from an aerial photograph the distortion caused by *tilt,* which occurs when the optic axis of the camera is not truly perpendicular to the horizontal plane.

Reflectance: the ratio of the radiant energy reflected by an object to the radiant energy received.

Refresh Tube: a CRT display on which the image disappears from the screen unless it is "refreshed" by a new pass of the electron beam about 30 or more times a second.

Replacement Operation: an arithmetic operation in which another, newly computed value replaces the current value of a variable.

Reproduction Separates: the sheets, or *flaps,* of film containing positive or negative artwork for various groups of map features used in the production of press plates for multicolor printing.

Rescaling: the adjustment of symbols representing magnitudes or intensity so that the visual appearance of the symbols will reflect the data values more closely than if simple proportional scaling were used; also called *apparent value rescaling.*

Residual: in trend surface analysis, the difference between the actual and trend-surface elevations of a point.

Right-reading Image: an image not reversed, that is, not a mirror image.

ROM: an acronym for read-only memory, a microcircuit containing programs or data that cannot be erased. When new data or programs can replace old ones, the microcircuit is called an EROM, for erasable read-only memory, or a PROM, for programmable read-only memory.

Scan Line: (1) one of the parallel tracks covered by the electron beam of a CRT display or the oscillating mirror of an airborne or satellite sensor; (2) a simple row of symbols on an image produced by a line printer, matrix printer, or a similar device.

Scratch File: a temporary file, used by a program for intermediate calculations and erased or released after computation is completed.

Scribecoat: a transparent film coated with a photographically opaque base that can be cut with specially designed tools to yield a negative image of linework or point symbols.

Search Region: the area around a point containing all data points to be used in estimating surface elevation at that point. A circular search region may be defined by its *search radius.*

Sequential Access File: a file, such as a magnetic tape file, for which the time required for access is dependent upon the location of the information most recently stored or retrieved.

Servomechanism: a control system the operation of which is modified by feedback so that the difference between actual and intended operation is reduced.

Sine: the ratio between the side opposite a specified acute angle in a right triangle and the hypotenuse.

Smart Terminal: a terminal with a minicomputer able to perform its own computations as well as provide communication with a host computer.

Soft Copy: a temporary image of a map or diagram, for example, on the screen of a CRT display.

Software: programs and data files.

Spectral Signature: numerical measurements of the properties of an object at a series of wavelength intervals.

Spline Interpolation: the interpolation of a smooth curve such as a topographic contour by fitting simple curves represented by mathematical formulas to portions of the overall curve in a manner providing smooth transitions between adjoining sections.

Stack: a linear list data structure with one "top" end at which data items are stored or retrieved.

Stereomate: an image accompanying an orthophoto and designed to yield a three-dimensional effect when the two images are viewed in stereo.

Stereoscopic Vision: the sensing of a single three-dimensional image by binocular vision of two perspective images generated from different vantage points.

Storage Tube: a CRT display on which an image can be stored on the screen for several minutes or longer with a single pass of the electron beam.

Subroutine: a programmed routine that can be "called" from the main program, or some other subroutine, at a point to which the computer returns after executing the subroutine; also called a *subprogram*.

Tabular Accuracy Index: a numerical measure of the internal homogeneity of a data classification for a choropleth map.

Tangent: the ratio between the side opposite a specified acute angle in a right triangle and the adjacent side.

Terminal: a device for communicating with a computer, and usually including a keyboard and either a CRT display or printer.

Thiessen Polygon: a polygon including a single data point and bounded by line segments equidistant between the included point and nearby data points. All sites within the polygon lie closer to the included point than to any other data point.

Thinning: process whereby a linear feature is represented in a grid by a continuous series of cells each of which touches along its sides and corners no more than two other cells belonging to the feature.

Time Sharing: serving simultaneously two or more programs, each of which receives, in a brief turn, the attention of the CPU.

Topological: referring to such properties of geometric figures as adjacency that are not altered by distortion as long as the surface is not torn.

Tract: a small area, usually containing about 4,000 residents, for reporting the results of the federal census.

Training Data: a set of pixels already assigned to a distinct land-use or land-cover category and used to "train" a classification algorithm for remotely sensed data so that other pixels might be assigned to an appropriate category.

Translation: the relocation of a figure while maintaining the same angular orientation to the system of coordinate axes.

Tree Structure: the ordering of data items in a hierarchical list whereby each item is linked directly to one item above it and to several items below it.

Trend Surface: a relatively simple surface represented by a mathematical formula and intended to represent the broad, general spatial trends in a set of three-dimensional data.

Turing Machine: a theoretical concept, developed by A. M. Turing, useful in proving whether or not a task can be performed by a computer system.

Turnkey: an adjective used to describe a computer system consisting of both hardware and software that is delivered ready for immediate operation.

Underflow: a condition in which a value is too small to be represented in a word in memory, for example, when a program attempts to store the computed result 0.1×10^{-105} in a word not able to accept a negative exponent below -63.

Union: the set containing all objects belonging to one or the other, or both, of two sets.

UTM Grid: the Universal Transverse Mercator grid, a system of plane coordinates based upon 60 north-south trending zones, each 6 degrees of longitude wide, that circle the globe.

Value: degree of lightness or darkness of a color.

Variable: one or more words in memory manipulated by the program and identified by a label such as X, POPULATION, or J.

Vector: a directed line segment, which can be represented by the coordinates for the pair of end points. *Vector data* refers to data in the form of a list or lists of point coordinates.

Viewport: a rectangular frame with a specified size and location on the screen of an interactive graphics system, and within which a rectangular portion, or *window,* of the map is displayed.

Virtual Memory: a "virtually unlimited" memory obtained by dividing a program and its associated data into segments called *pages,* only some of which reside in internal memory at any time. When a particular page not in main memory is needed, it exchanges places with a page in internal memory.

Voice Decoder: a device that converts spoken words and numerals into digital data, and which thus is useful in digitizing for recording feature codes called out by the operator.

Voronoi Polygon: a polygon including a single data point and bounded by line segments equidistant from the included point and nearby points so that all sites within the polygon lie closer to the included point than to any other point; also called a *Thiessen polygon.*

Window: a rectangular portion of a larger mapped area selected for display, for example, a rectangle surrounding Poland and ignoring most other areas in a data base covering eastern Europe.

Windowing: the display of only those portions of a digital map lying within a specified frame or *window.*

Word: a group of bits or bytes that can be stored or retrieved as a unit.

Word Size: the number of bits or bytes assigned to a word, thereby determining the accuracy of decimal numbers and the maximum absolute value of integers that can be stored in the word.

Wrong-reading image: an image reversed from one reading from right to left to one reading from left to right, as in a mirror image.

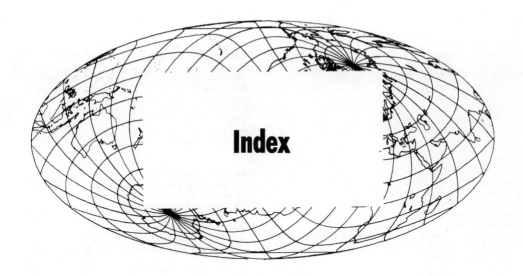

Index

Accessing, 141
Accumulating a sum, 27–29
Acoustic coupler, 146, 191
Acronym, 191
Address coding guide (ACG), 148–49, 191
  ACG/DIME files, 148–49, 156
Address in memory, 42–43, 138, 191, 194
ALGOL, 32
Algorithm, 30, 32, 191
Alphanumeric characters, 4, 37, 49, 191
American Cartographic Association, 186
American Congress on Surveying and Mapping, 186–87, 190
Anaglyph, 126, 192
Analog computer, 29, 192
Analytical solution, 54, 192
Angles, measurement of, 114–15
Animated map, 9, 192
APL, 32, 36, 92
Apparent-value rescaling, 91, 204
Arcs, 151–52, 192
Area, measurement of, 114–16
Area-fill algorithm:
  raster-mode, 70–73
  vector-mode, 104–106
Areal aggregation, 73, 158
Arithmetic logic unit (ALU), 23, 192
ARITHMICON, 151
Array processor, 43, 75, 192
Arrays, representation of in memory, 27–28, 43, 138
Artificial intelligence, 29–31, 176–77
Aspect, 76–78, 192

Assembler, 33, 192
Associated triangles, 155
Association for Computing Machinery, 189
Associative storage, 43, 192
Atlas layout, 159–61
Azimuth, measurement of, 114–15, 192

Band, spectral, 83–84, 192
Barriers to interpolation, 55
BASIC, 32, 92
Basic operations in data manipulation, 141
Batch processing, 22, 32, 34, 192
Bearing, measurement of, 114–15, 192
Binary digit (*see* Bit)
Bit, 25–27, 192
Bit-plane coding, 74–75, 192
Block, 138, 193
Block diagram, 130–36
Boundary segments, 142–45, 193
Bounding, 150–51
Brassel, Kurt, 19–20, 47, 80–81, 178, 181
Bug, 33, 193
Built-in functions, 33
Bulk storage unit, 24, 193
Byte, 25–27, 193

Cadastre, 21, 193
CAM (*See* Cartographic Automatic Mapping system)
Canadian Geographic Information System, 154
Canned program, 18–21, 33, 193